# Neurons *in* Action

## Computer Simulations with

John W. Moore • Ann E. Stuart

**Sinauer Associates, Inc., Publishers**
23 Plumtree Road, P.O. Box 407, Sunderland, MA 01375-0407

**Neurons in Action**
*Computer Simulations with NeuroLab*

Sinauer Associates, Inc. Publishers
23 Plumtree Road
Sunderland, Massachusetts 01375-0407 U.S.A.

Fax: 413-549-1118
Email: publish@sinauer.com

ISBN: 0-87893-537-1

Printed in U.S.A.

5 4 3 2 1

*To you, Jonathan,*
*our unflagging and good-humored colleague and teacher.*
*Without you it would not have been finished.*

# Preface

The purpose of *Neurons in Action* is to provide students with tools with which they can appreciate the complexity of the functioning of a single neuron. Students can perform unlimited virtual experiments on digital neurons to test and strengthen their understanding of neurophysiology.

The foundation of *Neurons in Action* is the extraordinary set of equations developed by Alan Hodgkin and Andrew Huxley in the 1950's to describe the results of their monumental experiments on the squid's giant axon. These equations remain the reference standard for describing the behavior of excitable membranes. Their unparalled accuracy allows computer simulations to predict nerve function under the wide variety of circumstances encompassed by these tutorials. Thus, the simulations in *Neurons in Action* reproduce the results of real experiments with remarkable fidelity.

An historical account of the conceptual, technical, experimental, and computational breakthroughs of these pioneers appears in the appendices of this volume; we are grateful for constructive criticisms and verification of its accuracy by both Hodgkin and Huxley. We are particularly appreciative of the permission to reproduce their papers on the *Neurons in Action* CD granted by Sir Andrew Huxley and the late Sir Alan Hodgkin's family.

# Table of Contents

# Setup *(one time only)*

## For Windows

- Check that your <u>hardware and software requirements</u> are met.

- <u>Set the resolution</u> of your monitor to 1024 x 768.

- Set up <u>Internet Explorer</u> or <u>Netscape</u> to launch NEURON simulations, then return to the <u>Home</u> Page and click "start NeuroLab".

## For Macintosh

- Check that your <u>hardware and software requirements</u> are met.

- <u>Set the resolution</u> of your monitor to 1024 x 768.

- Set up <u>Netscape</u> to launch NEURON simulations, then return to the <u>Home</u> Page and click "start NeuroLab".

# *Setup (one time only)*     *For Windows*

## Check that your Windows hardware and software requirements are met.

### Hardware requirements

- IBM-compatible PC
- CPU: Intel 486, equivalent or later
- Ram: 8 MB minimum
- Drive: CD-ROM
- Video card and monitor resolution: XVGA (1024 x 768)

### Software requirements

- **Internet browser**
  An Internet browser is required to view the NeuroLab tutorials, which are written in HTML. Internet Explorer is normally installed along with Windows 95 and 98. Netscape 4 is included on this CD for your convenience and possible installation.

- **NEURON**
  NEURON, NeuroLab's simulation tool (launched by the browser via hyperlinks), is on the CD. You must prepare Internet Explorer or Netscape to launch NEURON or the simulations will not run.

  For faster loading of the simulations, it is recommended you copy NEURON to your hard drive by dragging the "NRN" folder (~40 MB) from the root directory of the CD to the root directory of your "C" drive, using your file manager.

- **Acrobat Reader**
  Acrobat Reader is required to view the original papers that have been put on the CD in the PDF format. If you do not already have the Acrobat Reader installed, you may install version 4.05 from the CD as follows:
  1. Insert the Neurons in Action CD in your CD drive.
  2. Launch your file manager.
  3. Open the Acrobat.PC folder on the CD drive.
  4. Double-click on AR405ENG.EXE and follow the instructions.
  Alternatively, you may download the latest version of Acrobat from Adobe on the internet.

## Set the resolution of your PC monitor to 1024 x 768.

NeuroLab simulations have been designed for this resolution (also called XVGA). There are many different video drivers in PCs; consequently, NeuroLab simulations carried out on different machines reveal considerable variability in the location and size of the panels and windows on the computer screen. You will probably find that, even with the 1024 x 768 resolution, movement of the panels (whose size is fixed) and resizing of windows may be required.

While it is possible to work with a resolution of 800 x 600, you will have several difficulties at this resolution:

- Panels or windows that are located off screen cannot be easily found.
- Panels and windows can be inconvenient to move and size.
- Panels and windows may overlay and obscure each other.

If, because of your computer limitations, you can only increase your screen's resolution to 800 x 600 there is a makeshift help for you. Use the F1 function key to bring up a special control panel to allow you to rescale all panels and windows to fit your screen. You will find that this will work but may reduce the legibility of the labels and buttons.

### To set resolution:

1. Click on the "START" button.
2. Choose "Settings".
3. Choose "Control Panel".
4. Choose "Display".
5. Choose the "settings" tab.
6. In the Desktop area, position the slider to the 1024 resolution.
7. Click the "OK" button.

## Setting up Internet Explorer or Netscape browser

Attention! *The steps below MUST be completed in order for the simulations to run.*

### Set up Internet Explorer to launch NEURON simulations.

The Windows operating system needs to be told that any file with an "nrc" suffix should cause NEURON to run the simulations in the tutorials.

1. Double click on your "My Computer" icon.
   A window opens to display your computer's drives.

2. Double click on the CD icon labeled "Neurons in...."
   A window opens to display the contents of the CD.

3. Double click on the file named "Setup.nrc".
   An "Open With" window will appear, listing all of the programs currently recognized by Windows.

4. Click "Other" to prepare to register NEURON in this list.
   A new type of "Open With ..." window will appear.

   It allows you to tell the system where to find neuron.exe, the file that launches NEURON.

**5.** In the lower part of this new window type the location of the NEURON.EXE file.

If you have copied NEURON (the NRN folder) to your hard drive, type in the text exactly as shown below. If you prefer to run NEURON from the CD-ROM, replace the "C" with "X", where the "X" stands for the CD-ROM drive letter.

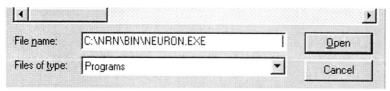

**6.** Click the "Open" button in this window

**7.** The original "Open With" window will reappear, now with NEURON and its icon added to the list of programs recognized by Windows.

The location that you typed in for finding NEURON.EXE will be shown in the white space near the top.

Make sure that the box associated with "Always use this program to open this file" is checked.

Click the "OK" button on the "Open With" window to confirm that this is the location where Windows should find NEURON.EXE

## Set up Netscape 4 to launch NEURON simulations.

**1.** Select "Preferences" from the browser's Edit menu.

**2.** In the window that appears (shown here), choose "Applications" under "Navigator" in the left-hand scrolling column called "Category".

**3.** Click the "New Type . . . " button to create a new application type. A "New Type" window, shown below, will appear.

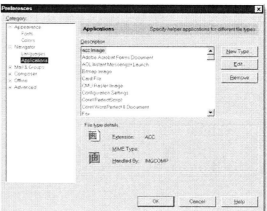

**4.** In the window that appears, fill in the blank fields, as shown at right, where the "D:\" stands for the CD-ROM drive. If the CD-ROM drive letter on your machine differs, use the appropriate drive letter.

If you have copied NEURON (the NRN folder) to your hard drive, use "C:" instead of "D:".

**5. Make sure there is a space before "%1" in the "Application to Use" field.**

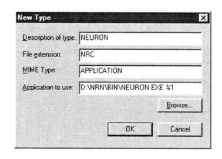

**6.** Click "OK" on the New Type and Preferences windows and return to the Neurons in Action home page.

# Setup *(one time only)*  ·  *For Macintosh*

## Check that your Macintosh hardware and software requirements are met.

### Hardware requirements

- A Macintosh with a G3 or PowerPC processor
- MAC OS 7.x or later
- Ram: 8 MB minimum
- Drive: CD-ROM
- Monitor resolution: XVGA (1024 x 768) is recommended.

### Software requirements

- **NEURON**
  NEURON, NeuroLab's simulation tool (which is launched via hyperlinks by the browser) cannot be run directly from the CD.

| Attention! | *You must copy NEURON to your hard drive by dragging the folder "NRN.MAC" from the CD window to your hard drive.* |
|---|---|

- **Internet browser**
  An Internet browser is required to view the NeuroLab tutorials, which are written in HTML. Internet Explorer is not recommended due to incompatibilities that cause computer crashes.

  Netscape 4.74 is included on this CD for your convenience and possible installation.

- **Acrobat Reader**
  Acrobat Reader is required to view the original papers that have been put on the CD in the PDF format. If you do not already have the Acrobat Reader installed, you may install version 4.05 from the CD as follows:
  1. Double-click on the Neurons in Action CD.
  2. Double-click on the folder titled "Acrobat.MAC."
  3. Double-click on "Reader Installer" and follow the instructions.
  Alternatively, you may download the latest version of Acrobat from Adobe on the internet.

# Set the resolution of your Mac monitor to 1024 x 768.

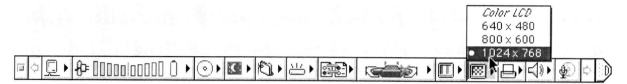

If you have System 8 or above or a Powerbook, you should have a Control Strip at the bottom of your screen, in which case you can set your resolution simply by selecting "1024 x 768" under the screen resolution icon.

Otherwise . . .

1. Select "Monitors & Sound" under "Control Panels" from the Apple menu. *(If you are using system 7.x or 9, the control panel is simply called "Monitors.")*

2. Choose "1024 x 768" in the scrolling field under the word "Resolution." If there is more than one "1024 x 768" choice on the list, try choosing "Recommended" in the popup menu, or experiment to find the selection that is best for your screen.

   **If "1024 x 768" is not an option, you may not have enough VRAM or a monitor that supports this resolution.** In this case, read about **the problems** you may encounter at lower screen resolutions.

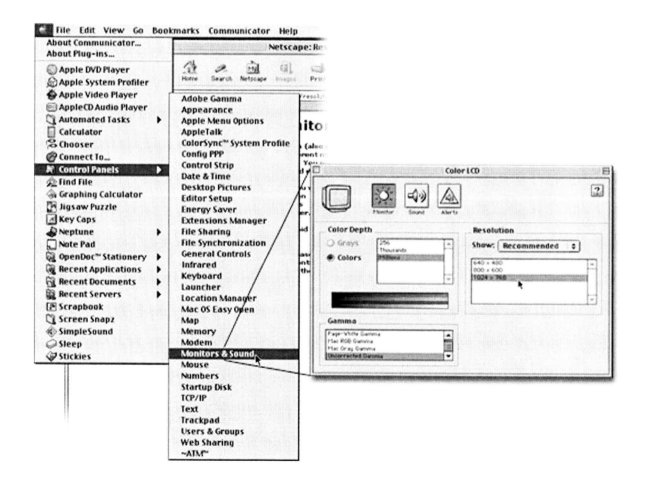

# Set up Netscape to launch NEURON simulations.

### Netscape
Netscape is available on the Neurons in Action CD under the folder "Netscape.MAC". If you would like to use Netscape, drag this folder to your hard drive and instruct it to run NeuroLab simulations as shown below.

## Instruct your Netscape 4 browser how to launch NEURON simulations.

**1.** Select "Preferences" from the Edit menu.

**2.** In the window that appears, choose "Applications" under "Navigator" in the left-hand scrolling column called "Category".

**3.** Click the "New..." button to create a new application type.

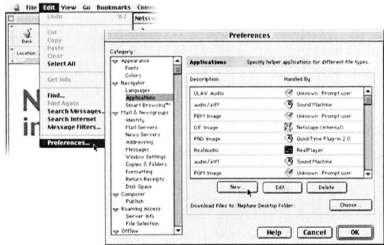

**4.** In the window that appears, type "NEU-RON," "application," and "nrc" in the three fields, as shown at right.

**5.** Click the radio button next to "Application" in the "Handled by" box. Then click "Choose...".

**6.** In the dialog box, locate your copy of Neuron (which MUST be on your hard drive).

**7.** Select "neuron" and click "Open".

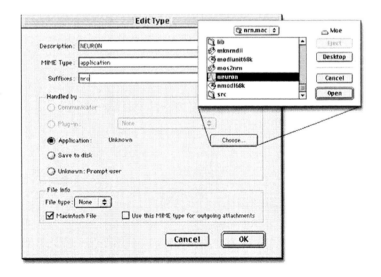

**8.** Click "OK" on the Edit Type and Preferences windows, return to the Neurons in Action home page and click "start NeuroLab".

# The Membrane Tutorial

How do the component parts of membranes (the capacitance of the lipids, the pumps, and the conductances of the channels) work together to permit voltage signaling? The simulations below approach this question by starting with the bare lipid bilayer membrane and adding new mechanisms in a stepwise fashion.

You will deliver a pulse of current across the membrane and observe how it changes the voltage across the capacitance of the membrane. Further, you will observe the capacitive current flowing onto the membrane during the pulse. These experiments are carried out in a short, isopotential segment of axon called a "patch."

*axon patch*

*patch pipette*

## Goals of this tutorial:

- To become familiar with NeuroLab panels and windows
- To experiment with applying charge to a membrane and recording the resulting transmembrane voltage and capacitive currents when the membrane:
    - is only a plain lipid biplayer
    - has only a Na/K pump that establishes a resting potential
    - has, in addition to the pump, a voltage-insensitive, non-selective leakage conductance
    - has, in addition to the pump and leakage channels, the voltage-sensitive Na and K channels described by Hodgkin and Huxley

## Set up the panels and windows for the tutorial.

In this first tutorial, the panels and graphs are described in considerable detail. Subsequent tutorials will assume that you have learned these basic NeuroLab manipulations. HELP is always available.

1. **Starting the tutorial**

    Clicking the button pictured here will start the tutorial by bringing up three panels on the left side of the monitor.

    > **Start the Simulation**

    Two control panels will appear in the upper-left corner of the monitor (overlying the Print and File Window Manager, a panel you can ignore for now). A stimulus control panel will appear below the two control panels. The functions of these panels are described here; more details may be found by clicking on each panel's link.

2. **Calling up graphs and running simulations**

    *a. Bring up graphs and other panels with the Panel and Graph Manager (P&G Manager).*

This panel appears here and in all of the tutorials in the upper-left corner. In this first tutorial you will press the buttons in sequence to bring up new Voltage-vs-Time graphs as the tutorial progresses.

*b. Run simulations with the <u>Run Control panel</u>.*

This panel appears here and in all of the tutorials to the right of the P&G Manager. Its buttons control the initiation and timing of the simulations.
- The "Reset" button allows you to set the membrane potential to the value in the white field. In this Membrane Tutorial the value begins at zero.
- The "Reset & Run" button starts a simulation.
- The "Stop" button stops a simulation.
- The "time (ms)" button shows the elapsed time as the simulation progresses.
- The "Total # (ms)" button allows you to change the total number of milliseconds in the simulation from the default setting. If you make a change, the time-axis of all graphs will also change.

## 3. Toggling between the text and the array of panels and graphs
The set of panels and graphs cannot be minimized. In order to return to the tutorial from the panels and graphs it is wise to keep a small piece of the text window showing so that you can click on it. Similarly, return to the collection of panels and graphs by clicking on any of them (but in an area that is not a button). You can also Alt-Tab between the text and the collection of panels and graphs.

## 4. Inserting a stimulating electrode
When the <u>Stimulus Control panel</u> is open, a stimulating electrode is inserted into the cell.

*a. To deliver a current pulse, the Stimulus Control panel MUST be present (open).*

Closing the panel removes the electrode. If you fail to observe a current pulse or voltage change on a graph when you press Reset & Run, you probably have this panel closed by accident. You will have to press the "Restore Stimulus Control" button to reopen it.

The line in the center of the large white field in the bottom portion of this panel indicates where the electrode is located. In this Membrane Tutorial, the line represents an isopotential neuronal <u>patch</u> (equivalent to a soma).

*b. Note the pulse parameters in this panel.*

If you wish to see (or, later, change) the pulse delay, duration, or amplitude, click the "Iclamp" diamond in the upper portion of the panel. The pulse parameters will appear in the lower portion. Click the "Location" diamond to return to the representation of the electrode in the patch.

## 5. Recording voltage
It is not necessary to insert a recording electrode into your cell. The voltage-recording electrode is automatically inserted when you call up a Voltage-vs-Time graph (or, later, a Voltage-vs-Space graph) as you progress through the tutorial. Consult the <u>Plotting Windows</u> link for more details about the NeuroLab graphs.

### 6. Calling up the <u>Membrane Parameters</u> panel

Press the "Membrane Parameters" button to call up the panel that shows the size of the patch in this tutorial and the values of the parameters of the patch. Note the following:

- The patch area is 0.0001 cm$^2$ ($10^4$ μm$^2$), similar to that for a spherical soma of about 55 μm diameter.
- The membrane's capacitance is 1 μF/cm$^2$, the usual value for biological membranes.
- At the beginning of the tutorial all of the conductances are set to zero.

You can do experiments by <u>changing the values</u> in this panel. If you change a value, remember to change it back to the default setting before going to the next step of the tutorial.

 *To close a panel or graph, click on the word "close" at upper left, NOT on the standard close box in the upper-right corner.*

## Experiments and observations

The lipid bilayer membrane without its channels is simply a capacitor (C) ready to store charge. Charge (Q) stored on the membrane gives rise to a voltage (V) across it (the membrane potential). <u>As with any storage device</u>, the rate at which the membrane may be charged depends on its size (capacitance) and on the rate of flow of charge (current) into it.

Once the capacitor has been charged to a certain level of voltage (V = Q/C), the charge will remain in place and the voltage will be maintained indefinitely.

### *Experiment with charging the capacitance of a lipid bilayer with a current pulse.*

### 1. First observe how a current pulse charges a membrane.

*a. Press the "Plain Bilayer Membrane" button (in the P&G Manager) to call up a Voltage-vs-Time graph.*

The graph appears in the upper right-hand corner. In this simple first experiment you will deliver a current pulse to the membrane and observe the rate of change of the voltage across the membrane. Note that depolarizations will be plotted from zero: as yet there is no resting potential in this cell.

*b. Press the "Capac. Current-vs-Time" button to call up a second graph.*

A graph will appear beneath the Voltage-vs-Time graph on which the capacitive current density (capacitive current per unit area) will be plotted when you run the simulation. Since the current pulse will be 2 nA, and the area of the patch is 0.0001 cm$^2$ (displayed on the Membrane Parameters panel), the capacitive current density across the plain lipid bipayer will be 0.02 mA/cm$^2$.

*c. Run the simulation: Charge the membrane.*

When you press the "Reset & Run" button in the Run Control panel the stimulating electrode will deliver the current pulse to the membrane, the voltage across the membrane will be plotted, and the capacitive current flowing across the membrane will be plotted. The current pulse is on the graph only as a marker (there is no axis on this graph for units of current so its amplitude has no meaning). Observe that the membrane potential changes as a linear ramp until the current ceases; from that time forth the voltage is held at that level. The membrane is charged to a certain voltage.

d. *Pause to make measurements or proceed to the next step.*

You can measure the capacitive current and the rate of change of the voltage to see how they are related (or to check the accuracy of the simulator!). Consult the link.

e. *Do you understand capacity currents?*

Understanding capacitive current is the first step in understanding the famous Hodgkin-Huxley (HH) equation used throughout the NeuroLab tutorials. The first term in this equation is the capacitive current: the product of the membrane capacitance and the rate of change of the voltage. For the plain lipid bilayer, the HH equation is reduced simply to this first term because there is neither a leak conductance nor voltage-sensitive channels in this basic membrane.

f. *Store your traces.*

To save a plot for comparison with a new plot when you make changes below, use the "Keep Lines" feature. Place your cursor in the plotting window, click the right mouse button, hold it down, and then select this option. The cursor MUST be placed within the plotting window for this option to work.

2. **Change the amplitude of the current pulse.**

a. *Press the "IClamp" diamond in the Stimulus Control panel.*

The parameters of the current pulse will be revealed in the lower half of the panel. Change the current stimulus amplitude with the arrows to the right of the white value field or with the field editor. You MUST have the cursor positioned in the white field when you enter changes using the keyboard.

b. *Observe the change in voltage and capacitive current for the new pulse.*

Is the slope of the ramp directly proportional to the current? Should it be? Click this link if you are unsure.

If the plot axes are too large or need expansion, you can resize them by right-clicking on the graph and selecting View=Plot. If you wish to restore the original gain, however, you will have to close and reopen the window.

c. *Erase your plots and restore the default current setting to prepare for the next experiment.*

You can clear the plot at any time with the "Erase" feature in the menu called up by the right mouse button. A simpler way to erase is simply to bring up a new plotting window with the P&G Manager.

### 3. Change the membrane capacitance.

*a. Call up the Membrane Parameters menu.*

You may already have done this. If you have not, press the appropriate button in the P&G Manager. Change the capacitance and rerun the simulation. How is the slope of the ramp <u>related</u> to the capacitance?

*b. Restore the membrane capacitance to its default value.*

Always restore default values when you are finished with an experiment. You do this by clicking on the box with the red check mark on it.

## *Electrify the membrane (establish a resting potential) by adding the Na/K pump.*

In order for a nerve to send signals by an electrical mechanism, there must be an electrical field across the membrane. A charge separation of tens of millivolts across the nerve membrane is accomplished by a special ionic pump, the <u>Na/K pump</u> (or Na/K ATPase), because it simultaneously moves Na ions out of the cell and K ions in.

### 1. Press the "Add the Na/K pump" button to establish a resting potential.
The Voltage-vs-Time graph will be overlaid by a new one showing that the tutorial has simulated adding the pump by charging the membrane to a resting potential of -75 mV.

Note that the "Reset (mV)" button in the Run Control panel now has a red check mark and that its value has been changed from 0 to -75 mV.

### 2. Run the simulation.
Does the value of the resting potential <u>affect the slope</u> of the voltage ramp in response to a current pulse?

## *Add "leak" channels.*

Hodgkin and Huxley observed a voltage-insensitive (linear) current in the squid axon with a reversal potential near the resting value. They referred to this current as "<u>leak</u>" and incorporated it into their <u>equation</u> to enable their model, defined by this equation, to match their data. Here we will place these very simple "leak" channels into our plain membrane to see how adding a voltage-insensitive conductance affects the charging and discharging of the lipid bilayer capacitor.

### 1. Press the "Add leak channels" button.
A new Voltage-vs-Time graph will replace the previous one.

*a.* The value of the "leak" conductance assigned by Hodgkin & Huxley (0.0003 mho/cm$^2$) now shows up in the "Leakage cond." field of the Membrane Parameters panel. The present-day notation for conductance is "Siemens" (abbreviated S) rather than "mho" but NEURON uses the Hodgkin-Huxley term "mho" and consequently so do these tutorials.

*b.* The "Leakage potent." has been set to -70 mV to reflect the reduction in resting potential upon addition of a leak.

## 2. Run the simulation.
Now the <u>time course</u> of the change in voltage across the membrane is quite different from what you saw for a plain lipid bilayer.

In the <u>Hodgkin-Huxley (HH) equation</u> there are now two terms that describe the situation at this point in the tutorial: the first term, capacitive current, and the last term introducing the leak. The other two terms are still zero because there are as yet no voltage-sensitive channels in the membrane.

## 3. Be quantitative.
Measure the <u>amplitude and time constant</u> of the voltage change. These <u>measurements</u> are widely used in neurophysiological experiments.

## *Add Hodgkin-Huxley (HH) voltage-sensitive channels.*

Add the HH Na and K channels to the membrane to enable it to generate action potentials upon depolarization.

### 1. Press the "Add HH channels" button.
Again an appropriate Voltage-vs-Time graph will replace the previous one. The default values for the densities (conductances) of the Na and K channels (as well as the HH values for the leak conductance and reversal potential) will appear in the Parameters panel. The resting potential (viewed in the "Reset" field in the Run Control panel) is further reduced by the small leak of Na into the cell. This matter will be further explored in the <u>Resting Potential Tutorial</u>.

### 2. Run the simulation.
You should observe that two spikes are generated during the current pulse. The capactative current surges at the onset and offset of the pulse as it did in the previous exercise, but during the action potentials it has a very interesting shape. (Remember that you can "turn down the gain" for the capactive current plot using the View=plot option on the right mouse button pop-up menu; to restore the original gain, close and reopen the window.)

### 3. Consider these questions:
• Do you know at what point in the action potential the capacitive current is at its peak?
• At what point in the action potential does the capacitive current cross zero?
• Why is the capacitive current during the falling phase of the action potential more prolonged but smaller?
• Click here for <u>answers</u>.

**4. Experiment with increasing the duration of the current pulse.**
Increase the stimulus current duration and then increase the "Total# ms" in the Run Control panel as you did above. Will spikes <u>continue to be generated</u> at this rate indefinitely?

## *Summary*

The two composite figures below summarize the four steps of this tutorial. They show the voltage responses to the current pulses at low gain and then at high gain and expanded time base to enable you to resolve the differences between the traces in the first few milliseconds.

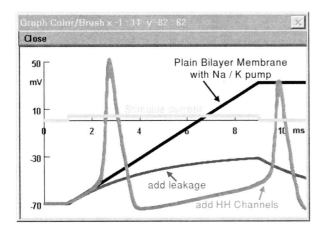

- **Black trace**: The voltage follows a linear ramp when charging the bare membrane.
- **Brown trace**: The addition of leakage channels causes the voltage to deviate from the ramp after an initial rise along the ramp's trajectory. The rate of the voltage change decreases as the voltage rises towards a steady-state level. At the end of the current pulse the voltage decays exponentially back to the resting potential.
- **Red trace**: The addition of voltage-sensitive Na and K (HH) channels does not cause a noticeable change (from the voltage rise when only leak channels are present) until the explosive, regenerative action potential arises. The voltage then quickly surpasses the linear ramp characteristic of the purely capacitive membrane.

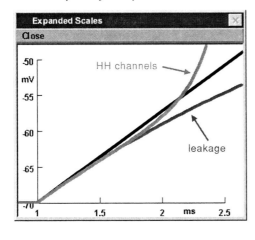

In this expanded version of the previous figure you can see that the the **red curve (HH channels)** initially overlies the **brown curve (leakage conductance)**. This happens because the resting state conductances of the HH channels add insignificantly to that of the leakage.

## *Press "Quit" to go to the next tutorial.*

1. **Press the "Quit" button (in the P&G Manager).**
   This button terminates the tutorial. It is distinct from "Close," which simply closes a panel.

2. **Confirm by clicking on the "Yes" button in the dialog box.**

*You MUST quit each simulation before going to the next to avoid having multiple copies of NEURON running, each with a different simulation and set of panels, electrodes, and settings.*

# Patch Resting Potential Tutorial

The resting potential is determined by the internal and external concentrations of the ions to which the membrane is permeable and obeys the Nernst equation. In this tutorial you can change ion concentrations inside and outside the axon patch, change membrane permeabilities, and plot membrane potential values to see if your values obey the Nernst equation.

*axon patch*

*patch pipette*

## Goals of this tutorial:

- To measure the membrane potential when the membrane is permeable only to a single ion species and explore its dependence on the external and internal concentration of that ion
- To measure the resting potential, the value of the membrane potential when the membrane has its normal ion conductances
- To compare your observations with published experimental results

<table>
<tr><td>

**Please Note!**

</td><td>

*All of the graphs in this tutorial will plot voltage as a function of time after a change in conductance or ion concentration has been made at t = 0. Because the properties of the Na and K channels are voltage and time-dependent, the voltage will undergo changes before settling down to a steady-state value. This tutorial focuses on the steady-state voltage.*

</td></tr>
</table>

## Set up the panels and windows for the tutorial.

**1. Assumptions**

This tutorial assumes that you are familiar with the basic manipulations of NeuroLab from The Membrane Tutorial. Descriptions of certain useful operations will be repeated in the text of this tutorial. Detailed descriptions of panels, windows, and manipulations may be accessed through the links in the tutorial.

**2. Starting the tutorial**

Click this button to launch the first four panels of the tutorial.

### Start the Simulation

Two control panels will come up in the upper-left corner:

- the Panel and Graph Manager (P&G Manager) with buttons specific for this tutorial
- the Run Control panel that controls the running of the simulations

These panels overlie the Print and File Window Manager that displays where your windows are located on the monitor screen. You can generally ignore this hidden window. A Patch Parameters panel comes up below the control panels.

### 3. The Patch Parameters panel

The <u>Patch Parameters panel</u> displays the parameters for this tutorial. You can eventually <u>change</u> any of the parameters in your experiments, but for the first steps of this tutorial they are programmed to change with the step.

- *Membrane conductances*
  The default values of conductance, assigned for squid axons by Hodgkin and Huxley, are proportional to the ionic channel densities. Later in the NeuroLab tutorial sequence the values of these conductances will have greater meaning. The channel conductances for Na, K, and <u>leak</u> are reset to experimental values when you are asked to press a button at each new tutorial step.

- *Ionic concentrations*
  These concentrations are typical for mammals, frogs, and other terrestrial animals.

- *Equilibrium potentials for the ions*
  The <u>equilibrium potentials</u> are calculated automatically from the Na and K concentrations set above.

### 4. The graphs

In this tutorial, a specific <u>graphical window</u> for plotting membrane potential versus time will come up in the upper-right corner of the monitor at each tutorial step. Launching this window inserts a recording electrode into a small patch of axon about the size of a node of Ranvier. (Be grateful for a simulator that takes the hassles out of experimentation!)

 **Attention!** *Single click on a button. Double clicking will bring up the panel or window twice!*

## Experiments and observations

*Observe the membrane potential when the membrane is permeable only to Na ions.*

1. **Press (click) the "Na Conductance Only" button in the P&G Manager.**
   A Voltage-vs-Time graph specific for this situation will appear in the upper-right corner. Note in the Patch Parameters panel that the Na conductance is not affected when you click this button but that the K channel and <u>leakage</u> conductances are automatically set to 0.

2. **Run the simulation.**
   Do this by clicking the "Reset & Run" button (R&R) on the <u>Run Control panel</u>. In the Membrane Voltage-vs-Time plot you will see three horizontal lines and their labels:

   - The **red line** indicates the membrane potential.
   - The **blue line** represents the equilibrium potential for Na, **ENa**.
   - The **brown line** represents the equilibrium potential for K, **EK**.

The membrane potential will change over the course of a few milliseconds and then settle at ENa. Check out whether the membrane potential is exactly at ENa by <u>measuring</u> the value with the crosshairs. Use this value to begin to make a plot of ENa vs log[Na]out (on paper). For a <u>discussion</u> of what determines the time course of the change in the voltage, click the link. You may wish to ignore this subject for now and concentrate on the steady state.

### 3. Change the external [Na].
In the Patch Parameters panel, change the extracellular [Na] from its default value of 140 mM to higher and lower values. For example, double the concentration, then decrease it by factors of two or four. Each time you change the concentration, run the simulation, measure the membrane potential with the crosshairs, then plot the value of the membrane potential versus the log [Na]out.

### 4. Question:
What is the <u>value</u> of the slope of the line through the points you have plotted? Is it similar to the second figure shown in <u>this link</u>?
*Hint:* Review the <u>Nernst equation.</u>

## *Observe the membrane potential when the membrane is permeable only to K ions.*

### 1. Press the "K conductance only" button.
A new graph will replace the previous one. In the Patch Parameters panel you will see that the K conductance is reset to its nomal Hodgkin-Huxley value and that the Na and leak conductances are set to 0.

### 2. Run the simulation.
Note that the membrane potential now changes in the negative direction from its previous value (-65 mV) until it (almost) overlies EK. Click the link for a <u>discussion</u> of the time course of the change in voltage. If you wish to bring the membrane potential to EK more quickly, increase the K channel density (K conductance) by one or two log units.

### 3. Measure the membrane potential in normal [K] and different [K]out.
As you did when you changed the [Na]out, make a plot of EK vs log[K]out. In the Patch Parameters panel, change the extracellular [K] from its default value of 5 mM to higher and lower values. Plot the value of the membrane potential versus the log [K]out. Compare the slope of the line through the points you have plotted to the second figure shown in <u>this link</u>.

## *Observe the membrane potential when the membrane is permeable to both Na and K.*

### 1. Press the "Na, K, and Leak" button.
Now the membrane has the conductances that determine its normal resting

potential. When you run the simulation the membrane potential should maintain its resting value of -65 mV.

## 2. Change the [K]out again.

Change the extracellular [K] from its default value of 5 mM to higher and lower values. Make one last plot the value of the membrane potential versus the log [K]out.

| Avoid | *Avoid [K]out in the 7 mM to 16 mM range where the membrane potential oscillates due to limitations of the Hodgkin-Huxley equations.* |
| --- | --- |

## 3. Questions:
- Why does the resting potential curve <u>deviate</u> from that for EK, which changes according to the Nernst equation?
- Can you <u>explain</u> the (sometimes very large) transient changes in the membrane potential on the way to the final value?
- Can one <u>increase the external K concentration</u> significantly without eliciting regenerative activity?

## 4. Change the [Na]out.

Return [K]out to its default value by clicking on the red check mark. Divide the default value of [Na]out (140 mM) by a factor of four or five for each point until you are in the vicinity of 0.5 mM. Measure and record the steady-state potentials for each [Na]out. You can do this by using the crosshairs as above.
- <u>Why</u> is the resting potential so insensitive to the Na concentration?

## 5. <u>Compare</u> your simulated results with classic observations.

# Now press "Quit" and go to the next tutorial.

Press the "Quit" button in the P&G Manager.

| Remember! | *You MUST quit each simulation before going to the next one to avoid having multiple copies of NEURON running, each with a different simulation and set of panels, electrodes, and settings.* |
| --- | --- |

# Patch Action Potential Tutorial

The action potential is a transient, regenerative voltage response provoked by a variety of stimuli. The non-propagating action potential in a patch (sometimes called the "membrane action potential") is easier to understand than the propagating action potential in an axon.

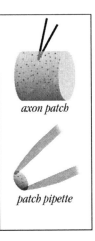

*axon patch*

*patch pipette*

In the Patch Resting Potential Tutorial you saw that a quick change in the bathing medium, one that depolarized the neuron, could produce a regenerative response. In this tutorial you will use the more common stimulus of injecting positive current into the neuron to trigger an action potential.

## Goals of this tutorial:

- To observe the action potential, and its underlying currents and conductance changes, in a uniform patch of membrane
- To experiment with how action potentials are affected by temperature, anesthetic agents, and toxins
- To change the [Na]out and observe the effect on the peak and timecourse of the action potential.

## Set up for the tutorial.

### 1. Assumptions

This tutorial assumes that you are now familiar with the following panels and manipulations:

- The function of the buttons within the <u>P&G Manager</u> and <u>Run Control panels</u>
- Running simulations by clicking Reset and Run (R&R) or clicking Reset and then the "Continue for (ms)" button in the Run Control panel
- Inserting stimulating electrodes by launching the <u>Stimulus Control</u> panel and controlling the onset, amplitude, and duration of the stimulus
- Storing, erasing, resizing, and measuring values on traces in <u>plotting windows</u> using the <u>right mouse button</u> features
- Using the <u>field editor or the arrows box</u> to change values in the <u>parameters panels</u>

If this tutorial is your introduction to Neurons in Action, we suggest that you familiarize yourself with the panels and operations listed here by clicking their links.

### 2. Starting the tutorial

## Start the Simulation

Clicking this button will bring up the P&G Manager and Run Control panels of the tutorial. Buttons on these panels will bring up the other panels and graphs.

### 3. Inserting a stimulating electrode

To insert the stimulating electrode, click the "Stimulus Control" button. Click the "IClamp" diamond in the upper portion of this panel to view or change the pulse parameters, shown in the lower portion. Click the "Location" diamond to return to a representation of the electrode.

### 4. The graphs

Graphs for recording voltage, currents, and conductances in plotting windows are launched by pressing the appropriate buttons in the P&G Manager. At this point press the "Voltage-vs-Time Plot" button to display a graph of the membrane potential. Other buttons in the P&G Manager will call up the other graphs as the tutorial progresses.

### 5. Running a simulation

Run a simulation by pressing Reset & Run (in the Run Control panel), referred to from now on as R&R.

If you had pressed R&R before pressing Stimulus Control, you would have seen a red horizontal line indicating the membrane potential measured by the recording electrode, which is always in the neuron. There would have been no action potential because you would not have inserted a stimulating electrode into the patch.

 *If you close the Stimulus Control panel, the electrode is removed from the patch. Also, do not select "Stimulus Control" more than once unless you intend to insert additional electrodes into the patch.*

## Experiments and observations

### *Generate action potentials.*

### 1. Press R&R to generate an action potential.

You will see an action potential displayed on the Voltage-vs-Time graph in red.
- The brief stimulating current pulse is shown in green.
- ENa is shown as a blue line.
- Ek is shown as a brown line.

### 2. Observe the currents underlying the action potential.

To do this, press Membrane Current Plots (in the P&G Manager) to bring up the appropriate graph and then run the simulation (press R&R). You will see graphs of the Na and K currents (patch INa and patch IK) that cause the rise and fall of the action potential.

If you are familiar with voltage clamp experiments, you will note that these action potential current patterns are not the same as those observed with a voltage clamp. The voltage clamp measures currents in response to the relatively simple stimulus of the voltage step (see <u>Patch Voltage Clamp Tutorial</u>); in contrast, the currents flowing during the complicated voltage change of an action potential cannot be measured experimentally but can only be calculated, as NEURON is doing here.

3. **Now observe the conductances changing during the action potential.**
Press Membrane Conductance Plots (in the P&G Manager) to bring up a graph for plotting the Na and K conductances (Patch gNa and Patch gK) as a function of time. Run the simulation.

Of course, you could not make these observations of conductance with a current clamp or even with a voltage clamp in a real experiment. They may only be calculated or seen using a simulator.

4. **Question:**
What underlies the <u>depolarizing ramp</u> at the beginning of the action potential?

## The shape of INa

Why is INa so "kinky" with two phases? Why does it not have a smooth time course since the voltage and gNa and even IK all have smooth time courses? Here are two experiments to assist your reasoning.

1. **Plot the driving force on the Na ions as a function of time.**
Press the "Plot Driving Force for INa" button (in the P&G Manager). An appropriate graph will come up, overlying the Voltage-vs-Time graph. Run the simulation by pressing R&R.
   • The red line is the action potential
   • The black line is the driving force on Na (Membrane Voltage minus ENa).

2. **Find the time at which the minimum (the notch) in INa occurs.**
Use the <u>Crosshairs</u> by clicking the left mouse button on the current trace; read off the time (the x value) in the blue bar at the top of the graph. Find the y values at that time on the curve that plots the INa driving force and on the plot of gNa. Can you now explain the <u>kink</u>?

3. **Measure the size of the peak INa.**
Keep a note of this value for comparison with Na currents in later exercises.

4. **Close the Driving force for INa-vs-Time graph.**
Close this panel to expose the Voltage-vs-Time graph beneath for the remainder of the experiments in this tutorial.

## More questions

• Does <u>changing either the length or diameter</u> of the patch alter the action potential in any respect? Should it?
• Does the <u>location of the stimulating electrode</u> in the patch matter? Should it?

- Suppose the lipids in the membrane were not such strong insulators and the membrane's thickness had to be doubled. What would this do the membrane's capacitance per unit area? Call up the <u>Patch Parameters</u> panel (in the P&G Manager) and change the value of the capacitance to see how your change would affect the shape of the action potential.

    When you change the capacitance, there will be changes in the rate of rise and in the final amplitude of the depolarization in response to the current pulse. Are these changes compatible with the explanation of <u>capacity charging</u> in the Membrane Tutorial?

- Can you find parameters that would permit you to elicit more than one action potential in this simulation?

## Study the effect of temperature on the action potential and underlying conductances.

### 1. Change the temperature (in Run Control).
You can warm or cool the patch by using the UP or DOWN arrows (to the right of the white value field) or by typing a new value into the field using the <u>field editor</u>.

### 2. Compare traces at different temperatures.
Use the "Keep Lines" option (<u>right mouse button menu</u>) for comparing traces. You must have your cursor positioned on the graph when you select "Keep Lines."

Note that your experiments so far have been carried out at 6.3°C, the standard reference temperature for squid, an invertebrate, used by Hodgkin and Huxley. Although the lack of heating in labs in post World War II England is the most likely reason for this low temperature in their experiments, the effect was to slow the changes in the ionic currents to the point where the electronic circuits could control the potential more accurately.

### 3. Draw conclusions from this important experiment.
What happens to the <u>duration</u> of the action potential if you change the temperature? Can you explain your observations by studying the effect of temperature on the underlying conductances? The results of this experiment are crucial for understanding why temperature affects the ability of an action potential to invade demyelinated regions of myelinated nerve (<u>Partial Demyelination Tutorial</u>).

## Bathe the axon in anesthetic agents.

### 1. Open the <u>Patch Parameters</u> panel so that you can do more experiments.
Do this by clicking on the "Patch Parameters" button (in the P&G Manager) if this panel is not already open.

### 2. Partially block both the Na and K conductances.
The anaesthetics procaine and lidocaine reduce both the Na and K conductances by almost equal factors. Reduce the values of the conductances (the Na channel density and the K channel density) by a factor of 2.

3. **Compare the normal and partially blocked action potentials.**
Using Keep Lines, compare action potentials generated at different "concentrations" of anaesthetic by continuing to divide by a factor of 2. By how much must you reduce the two conductances to block the generation of the action potential? Make a note of this value for comparison with experiments below using selective poisons of each channel type.

## Block the Na channels with the poison _tetrodotoxin_ or _saxitoxin_.

1. **Block the Na channels by reducing the Na conductance.**
Reduce the value of only the Na channel density by dividing by a factor of 2 as you did above until the regenerative response disappears. Use Keep Lines to compare the action potentials in normal saline and at different degrees of block of the Na channels.

2. **Question:**
Which is more effective at blocking action potentials, a toxin that selectively blocks Na channels or the anesthetics (investigated above) that block both Na and K channels? Why?

## Change the external and internal Na concentrations.

You can reproduce the experiments of _Hodgkin and Katz_ (1949) on squid giant axons. They altered the external Na concentration and found that the amplitude and rate of rise of the spike were tightly coupled to the changes in ENa. Furthermore, you can also do an experiment for which the techniques were not available at the time: you can change the internal Na concentration.

1. **Change the [Na]out.**
Make your changes over a wide range (e.g., by 2-fold steps) from double the default value down to a value equal to the internal concentration. Plot the peak of the action potential and ENa as a function of [Na]out. How do your results compare with those of Hodkgin and Katz?

   You can check your plot of _ENa versus [Na]out_ against ours.

2. **Change the [Na]in.**
Go beyond what Hodgkin and Katz were able to do and perfuse the axon internally with a solution of your choice. You can mimic a "very tired nerve" experiment by doubling the [Na]in and then repeating changes in the [Na]out. How should the plots change if the internal Na concentration is doubled?

# Patch Voltage Clamp Tutorial

The voltage clamp, conceived by Cole and used ingeniously by Hodgkin and Huxley, is a tool that has made possible extraordinary advances in our understanding of the basis of electrical activity in nerve cells.

*axon patch*

*patch pipette*

With the voltage clamp, Hodgkin and Huxley stepped the voltage across the membrane to values more positive than the resting potential and observed the currents flowing across the membrane. At these values the Na and K channels opened, each with their own time course, and currents flowed through the channels. From these currents Hodgkin and Huxley developed the equations, here called the HH equations, which allowed them to calculate the conductance changes giving rise to the action potential. You might find it helpful to read about their experiments in the Appendix material before proceeding with these simulations.

NEURON makes use of the HH equations to calculate how the Na and K conductances change as a function of time for a given voltage step. It also calculates the currents, essentially calculating the data that Hodgkin and Huxley observed.

## Goals of this tutorial:

- To plot currents in response to individual depolarizing voltage steps and to families of voltage steps
- To plot the conductance increases in response to these voltage steps that underlie the changes in current. They should allow you to appreciate how the opening and closing of the Na and K channels shape the action potential.
- To experiment with "tail currents" (that give information on the time course of the closing of the channels when the voltage is returned to rest)
- To use the voltage clamp to demonstrate that a portion of the Na channels are inactivated at rest
- To experiment with the effect of temperature on the kinetics of the conductances

## Set up the voltage step.

### 1. Assumptions
This tutorial assumes that you are familiar with the following panels and manipulations:
- The function of the buttons within the P&G Manager and Run Control panels
- Running simulations by clicking Reset and Run (R&R) or clicking Reset and then the "Continue for (ms)" button in the Run Control panel
- Inserting stimulating electrodes and controlling the onset, amplitude, and duration of the stimulus using the Stimulus Control panel
- Storing, erasing, resizing, and measuring values on traces in plotting windows using the right mouse button features
- Using the field editor or the arrows box to change values in the parameters panels

If this tutorial is your introduction to *Neurons in Action*, we suggest that you familiarize yourself with the panels and operations listed here by clicking their links.

**2. Starting the tutorial**
Click this button.

> **Start the Simulation**

A P&G Manager, with buttons appropriate for this tutorial, and the Run Control panel will come up.

**3. Inserting the current-passing electrode**
When you launch the Stimulus Control panel, it comes up in the IClamp mode by default. For this voltage clamp tutorial you must select VClamp in the upper section of the panel. When you click the VClamp diamond, a set of voltage clamp parameters will appear in the bottom half of the stimulus control panel specifying the amplitude and duration of the voltage before, during, and after the voltage step. The default settings specify this voltage pattern:
- The "Conditioning" (holding) value at the resting membrane potential, -65 mV
- The "Testing" value, a jump to 0 mV
- The "Return" value of -65 mV
In the exercises you will change these values.

**4. The voltage clamp step**
Call up the Membrane Voltage-vs-Time plot (P&G Manager). Press R&R and you will see that the voltage is stepped to a given level and then returned. (If you instead see an action potential, you did not choose VClamp in the Stimulus Control panel).

# Experiments and observations

*Observe the Na and K currents in response to a step depolarization.*

**1. Plot the currents in response to a voltage step to 0.**
Bring up the Current-vs-Time plotting window by pressing Membrane Current Plots (in the P&G Manager). Run the simulation. You will observe the Na current (patch.ina) and the K current (patch.ik) as a function of time for the default step from -65 mV to 0 mV and back.

**2. Measure the peak Na current.**
Use the Crosshair feature to make the measurements. How big is the Na current here? It is instructive to compare this Na current with the peak Na current underlying the patch action potential. You may remember that there were actually two peaks to that Na current (separated by the "kink"); the larger of these peaks was about 0.8 mA/cm$^2$. What accounts for the difference in the magnitude of the Na current in these two situations?

3. **Vary the Testing Level of the voltage pulse.**
   Change the value of the depolarizing test pulse (using the VClamp menu in the Stimulus Control panel) in increments of, say, 10 mV. Find out how the peak INa varies with the level to which the voltage is clamped. Get out an old-fashioned piece of graph paper and plot the peak INa versus Vm, then compare your plot with the classic figure from the work of <u>Moore and Cole</u> (1960).

---

## Observe the Na and K conductance changes in response to a step depolarization.

1. **Bring up the Membrane Conductance Plots window.**
   You can observe how the Na and K conductances change as a function of time during the step to 0 mV. Conductances must be calculated—they cannot be measured. NEURON calculates each conductance from the rate equations and then the ionic currents from the driving force on the ion using <u>Ohm's law</u>, just as Hodgkin and Huxley did.

2. **Deliver the voltage step.**
   Press R&R. The Na conductance (patch.gna) and the K conductance (patch.gk) will be displayed on the same axis.

3. **You can delete the traces for one conductance to view the other set in isolation.**
   If you wish to plot only one conductance at a time, delete the other one using the Delete option in the right mouse button menu. With the cursor in the Conductance-vs-Time window, click Delete in the submenu. The words "Graph Delete" show up in the blue bar at the top of the graph, showing that the submenu is in the Delete mode. Click the *left* mouse button when the cursor is on the curve you want to delete.

*If you wish to delete the Na or K traces permanently from the plots, make sure the Delete feature is selected, then click on the text label for that trace. Rerun the simulation.*
*To replot the deleted traces, you will need to call up that window again in the P&G manager.*

4. **Explanations of the conductance kinetics are in the Appendix.**
   The kinetics of the conductances are not conventional. The description of the kinetics required the brilliance and insight of Hodgkin and Huxley. If you are eager to read about the kinetics at this point, you can pursue explanations of <u>the sigmoid lag in the onset</u> of each conductance and why there is <u>no sigmoid lag</u> when the membrane is repolarized to resting level. You may, however, wish to read this material in context in the narrative of the Appendix (or not at all!).

## *Observe families of currents.*

**1. Deliver 12 voltage steps from -40 mV to +80 mV.**
This experiment simulates a typical voltage clamp experiment (on a real neuron) where preset protocols are run by computers. You will deliver a <u>family of voltage steps</u> and observe a family of currents. In the process of generating the display of these currents, NEURON will also calculate the underlying conductances.

*a. Follow these steps to generate the family.*

This family by default will be a series of 12 steps in increments of +10 mV starting from -40 mV (holding potential -65 mV).
1. While still within the VClamp option (in the Stimulus Control window), set the Test Level amplitude to -40 mV for the initial voltage level.
2. Select VClamp Family. Notice that the "# of steps" button by default indicates 11, not 12. The first trace is the one for -40 mV and then there are 11 additional steps.
3. Select the "Keep Lines" option in each of the plot windows.
4. Press the "Vary Test Level" button, which is equivalent to pressing R&R.
5. Resize a plot with the ViewPlot option if necessary.

| **Attention!** | *You must press Vary Testing Level, NOT R&R, to generate a voltage clamp family. Run Control buttons apply only to individual runs.* |

*b. Observe the time courses and amplitudes of the Na and K currents.*

The Na currents at any level of depolarization precede the K currents. Where is threshold? If you can't answer this question, you are a good candidate for the <u>Patch Threshold Tutorial</u>.

The Na current family is rather complicated and will be sorted out by further experiments below. The K currents are easier to observe as a family. Notice the changes in the time course and amplitude with greater depolarizations. Observe how the K currents relate to the K conductance changes: the conductances go through a maximum change and then appear to saturate—that is, greater levels of depolarization cause no further increase in maximum conductance from the previous level. Yet the maximum currents continue to grow with level of depolarization. <u>Why</u>?

*c. Observe the Na currents in isolation.*

Delete the K traces by choosing the Delete option and clicking on the patch.ik and patch.gk labels in each graph. Click Keep Lines again in each graph. Run the simulation. Clearly the inward Na currents grow to a maximum as the level of depolarization increases, then decline and eventually become outward although the conductance increases steadily to a maximum. You can change the number of steps and the range over which they are delivered to observe the current at each voltage step more clearly.

*d. Consider these <u>questions</u> about the Na current.*

- Why do the inward currents grow larger with depolarization, then smaller?
- What is special about the voltage at which the Na current is zero?
- Why are the currents outward at large depolarizations?

*e. Prepare for the next experiment.*

Prepare for the next exercise by closing and reopening all the graphs for plotting currents and conductances so that the ability to plot K currents is restored. Return the Test Level pulse amplitude to its default value of 0 mV.

## *Observe "tail" currents.*

What is a tail current and what is its importance? Do this series of experiments to find out.

**1. Observe Na and K tail currents.**
Observe currents and conductances at the offset of a voltage step to 0 mV (upon its return from the Testing Level to the Return Level).

*a. Follow these steps to generate a tail current:*

1. Select VClamp in the Stimulus Control window.
2. Reduce the duration of the Test pulse to 1 ms. (Leave the Conditioning pulse at 0.5 ms and the Return pulse at 100 ms.)
3. Select Keep Lines in each graph to preserve these traces for the next experiment.
4. Press R&R (you are delivering a single pulse, not a family).

*b. Make your observations.*

Observe the direction, amplitude, and time course of the peculiar current, especially the Na current, at the end of the pulse. In addition to being at the tail end of the pulse, it really looks like a tail! You can increase the time resolution in either of these windows by setting the Total # ms (RunControl) at, say, 2 ms. Notice also how each of the conductances is changing when the tail currents are flowing.

**2. Determine how the tail currents depend on voltage.**
Clamp to three different Return Levels—say, -20 mV, 0 mV, and +20 mV—to see how the magnitudes of the currents depend on the voltage level. Again the behavior of the Na and the K currents is very different. Can you formulate an hypothesis to account for your observations and explain <u>tail currents</u> (before checking the link!)?

**3. Consider these questions:**
<u>What causes the Na tail current to jump</u> to a value that is larger than the value just before the the end of the pulse?

Can the amplitudes of the jumps in Na tail currents be used to <u>measure the time course of gNa</u>? Remember that gNa is not directly measurable but may only be calculated by dividing the current at any point by the driving force.

## Demonstrate inactivation of the Na conductance.

**1. How does inactivation of the Na channels depend on voltage?**
At the normal resting potential of -65 mV, about 40% of the HH Na channels are inactivated. Hyperpolarization removes inactivation with time; at a much greater resting potential, say -100 mV, none of the channels will be in the inactivated state. On the other hand, a depolarized holding (conditioning) potential will shift more of the channels to the inactivated state.

Note: The voltage level preceding a test pulse is standardly referred to as the "conditioning level." When the duration of this pulse becomes long enough so that the channels are in a steady state (about 3 ms at 6.3μ°C for the HH channels), the conditioning pulse may be referred to as the "holding potential."

**2. Set pulse parameters for observing inactivation.**
You can observe the consequences of inactivation by generating a family in which the Conditioning Level is changed but the Testing Level is kept constant. Adjust the parameters of the clamp step as follows:
1. Select VClamp and set the Conditioning Level amplitude to -100 mV.
2. Set the Test Level amplitude to 0 mV.
3. Select VClamp Family from the Stimulus Control window.
4. Set the # of steps to 11 (remember—there will actually be 12 traces).
5. Set the # mV/step to 5.
6. Make certain "Keep Lines" is selected for all of the graphs.

**3. Generate the family by pressing Vary Conditioning Level.**
A family of Na currents and a family of K currents will be plotted. You can expand the time axis by increasing the Total # (ms) to 2 ms. You may wish to run this simulation several times. Simply click the Erase option on each graph before a rerun. The Keep Lines option will remain selected.

**4. Look carefully at the Na and K conductances and currents.**
Should the current <u>patterns</u> during the pulse mirror the conductance patterns for each ion?

## Observe the effect of temperature on the Na and K conductance kinetics.

Cooling the preparation will cause the channels to open and close more slowly and, conversely, warming it will speed up the kinetics of the conductances.

**1. Restore default settings.**
Return all pulse parameters in VClamp and VClamp Family to their default settings. Erase the graphs but continue to Keep Lines. Restore the Total # (ms) to its default value. Run the simulation to deliver a pulse from -65 mV to 0 mV and back as at the beginning of this tutorial.

## 2. Change the temperature in increments of 10°C.

First set the temperature to 10°C to make the experiment easier. Then rerun the simulation at 20°C and 30°C. Observe how the change in temperature affects the kinetics of the conductances and therefore the currents.

Hodgkin and Huxley found a 3-fold change in the <u>rate constants</u> underlying both the Na and the K conductances (and currents as well) for each 10°C change. This important observation is at the heart of the interpretation of all of the experiments in these tutorials in which you change temperature.

# Patch Threshold Tutorial

## What is "threshold"?

Threshold is a very important concept. Whether a synaptic potential is above or below threshold determines whether or not a neuron fires, and thus whether a muscle moves or a thought exists. In <u>textbooks</u> it is often stated that a spike will be generated whenever the membrane exceeds a certain fixed voltage level. Is this true? The experiments of this tutorial should enable you to draw a conclusion. Grasping the details of threshold is especially important in understanding synaptic integration.

*axon patch*

*patch pipette*

## Goals of this tutorial:

- To decide whether threshold is at a fixed voltage in neurons
- To explore the precision of the threshold value for a current stimulus such as a synaptic current
- To discover what parameters of the voltage response do or do not affect threshold

## Set up for the tutorial.

### 1. Assumptions
This tutorial assumes that you are completely familiar by now with all of the manipulations of the NeuroLab tutorials. HELP is always available.

### 2. Starting the tutorial

**Start the Simulation**

The P&G Manager, Run Control, and Stimulus Control panels come up.

### 3. The stimulating electrode
Call up the parameters of the pulse (click IClamp) and note that the pulse has a delay of 0.5 ms, is 15 ms in duration (it will outlast the scales on the graphs because of the delay) and begins at an amplitude of 2.4 nA.

### 4. The graphs
Press the "Voltage vs Time Plots" button to bring up two graphical windows:
- A conventional full-scale plot displaying the voltage (from -70 mV to +50 mV) as a function of time
- An expanded voltage scale showing the threshold region (from -70 mV to -50 mV) at a higher resolution

It will be useful to Keep Lines in each of these graphs and to erase the stored traces between experiments.

## Experiments and observations

### Is there a critical voltage threshold?

**1. Give pulses just below and above threshold.**
Depolarize your patch with four current pulses just below and above threshold. Begin with the default pulse of 2.4 nA and then increase the amplitude (in the IClamp menu of the Stimulus Control panel) to the following values:
- 2.486 nA (just below threshold)
- 2.500 nA (just above threshold)
- 2.600 nA (well above threshold)

**2. Examine the traces to decide where threshold has occurred.**
<u>At what value</u> of voltage is the threshold of the response (the voltage change) for each trace? Is there a critical voltage threshold? Is there a <u>critical stimulus current threshold</u>?

### Does the stimulus threshold depend on pulse duration?

Stimuli for neurons range from brief synaptic currents lasting fractions of a millisecond to sustained currents of sensory receptors. Further, the duration of a synaptic current may be prolonged by modulatory factors or drugs. It is important to know how threshold is affected as the underlying current is prolonged—and why. This section explores how increasing the duration of the stimulus changes the amount of current needed to bring the membrane to threshold.

**1. What current amplitude is necessary to bring a brief stimulus to threshold?**
Must you change the current amplitude from the previous experiments? If so, by how much must you change it? Is the current threshold indeed a precise value, as it seems from the experiment above, or is it impossible to determine, like the voltage threshold? Will a very small difference (for example, due to a few transmitter quanta more or less) near threshold be consequential?

*a. Set the pulse duration at 0.1 ms.*

This duration is comparable to that of the synaptic input (the end plate potential) at the neuromuscular junction. You will deliver what neurophysiologists call a "short sharp shock."

*b. Find the threshold current.*

Find the pulse amplitude that is just above threshold for generating an impulse. Do you find this value surprising? What is the <u>explanation</u> of the results of this experiment?

*c. Find the difference in amplitude between a subthreshold and suprathreshold stimulus.*

Search for the threshold for this short pulse by successive bracketing between suprathreshold and subthreshold stimuli. What is the percent difference between current amplitudes that evoke "all" or "none" responses? Can you now answer the question posed above: Will a few quanta more or less make a difference in whether a synaptic potential near threshold causes a neuron to fire?

Notice how long the voltage stays depolarized before it finally returns to rest. The stimulus is only 0.1 ms yet the voltage change outlasts the stimulus by two orders of magnitude!

2. **Double the duration: Mimic the treatment for myasthenia gravis.**
Increase pulse duration from 0.1 to 0.2 ms. Set the current amplitude at its subthreshold value of 84.5 nA. By how much must you now <u>decrease the amplitude</u> to bring a pulse of this duration back to a subthreshold value? Keep track of the current values as you do this experiment.

From this experiment you can appreciate the treatment of the disease myasthenia gravis, an impairment of neuromuscular transmission due to a paucity of acetylcholine receptors. The disease is treated by drugs that prolong the action of acetylcholine—in other words, increase the duration of the synaptic current. This approach is quite effective since, as you observed, if you double the duration of the stimulus you halve the amount of current required to excite the cell.

3. **Increase duration drastically: Mimic sensory receptor currents.**
Observe thresholds when the neuron is stimulated with long depolarizing pulses (such as the currents generating sensory receptor potentials or slow synaptic potentials).

   a. *Lengthen the duration one to two orders of magnitude.*

   Change the duration of the depolarizing current pulse to 1, 5, 10, and 20 ms and find the amplitude of the threshold stimulus for each duration. To accommodate the 20 ms pulse you will need to change the time axis for this experiment by typing an appropriately larger number into the Total # (ms) field. If you survive the tediousness of this experiment you are unlikely to forget the result!
   • What <u>relation between duration and stimulus amplitude</u> do you find in this time range? Is it different from that for brief pulses?
   • Why is the <u>shape of the subthreshold stimulus</u> the same for the 10 ms and the 20 ms pulse?

   b. *Give a sustained stimulus.*

   Receptor potentials often last for seconds. Stimulate with a steady current for a time that is very long (e.g., 50 ms or longer) compared to the duration of the action potential. Increase the Total # (ms) accordingly. Can you get the neurons to give more than one action potential by increasing the amplitude of the current pulse?
   • Question: <u>What do you observe</u> and how do you explain your observations?

## *How is the peak amplitude of the action potential affected by nearness to threshold?*

**1. Give a series of pulses of increasing amplitude.**
Choose a pulse of long duration (10 ms is good) just below threshold and then increase the amplitude of the stimulus in a series of steps so that the action potential is generated earlier and earlier in the pulse.

**2. Observe the values of the peaks.**
<u>What happens </u>to the amplitude and why?

## *What is an "off response" and what is its threshold?*

There are many examples of synapses in the central nervous system where an action potential is evoked when a neuron is released from inhibition. This phenomenon is called "disinhibition" and the action potential generated is referred to as an "off response."

**1. Inject a hyperpolarizing current pulse to mimic inhibition.**
Set the pulse duration to 10 ms and the amplitude to -3 nA. Increase the Total # ms to 25 ms (you will see why when you run the simulation). Gradually increase the hyperpolarizing current until something interesting happens.

Can you <u>explain</u> your results?

**2. Compare thresholds of an off-response impulse with one generated by a depolarizing pulse.**
Store off-response traces generated just below and just above threshold. Then generate a similar set of traces using a depolarizing pulse (current settings of 2.5 and 2.6 nA will work for this 10 ms pulse). <u>What is different</u> about threshold in these two cases and why?

## *Observe the effect of temperature on threshold.*

Changes in temperature have a profound effect on impulse initiation and propagation and are clinically important, as you saw in the <u>Partial Demyelination Tutorial</u>. An <u>extensive discussion</u> of this subject, leading by links to the rate constants of the Hodgkin-Huxley equations, is available.

**1. Bring up the Conductance Plots window.**
This window will overlie the Voltage-vs-Time, expanded axis plot. Observing the conductances as you do this experiment should aid your understanding of the effect of changes in temperature on threshold.

**2. Change the temperature.**
Deliver a short pulse (0.1 ms) for this experiment. Set the pulse amplitude at 84.6 nA (just above threshold). Use Keep Lines to plot a family of curves for a series of temperatures. But before you do this exercise, make a guess as to what will happen.

# Patch Refractoriness Tutorial

## The threshold of an impulse changes following activity.

It is well known that there is a period of reduced excitability following an action potential called the refractory period. You have already encountered the refractory period in the Patch Threshold Tutorial where you observed that lingering changes in the ionic conductances following one impulse affected the threshold of the next impulse in a train.

*axon patch*

*patch pipette*

The experiments here explore how much additional current is necessary (for example, in a synaptic input) to overcome the increased conductance during the refractory period and generate a second impulse.

## Goals of this tutorial:

- To explore the duration of the period of refractoriness following an action potential
- To appreciate the amount of current needed to overcome this refractoriness at various times after an action potential
- To appreciate the ability of a subthreshold stimulus, such as an excitatory synaptic potential, to affect the subsequent excitability of a neuron
- To appreciate that even hyperpolarizing events, such as inhibitory synaptic potentials, can affect the subsequent excitability of a neuron

## Set up for the tutorial.

### 1. Assumptions

This tutorial assumes that you are completely familiar by now with all of the manipulations of the NeuroLab tutorials and that you have done the experiments in the Patch Threshold Tutorial.

### 2. Starting the tutorial

### Start the Simulation

The P&G Manager and the Run Control panel will come up.

### 3. The stimulating electrodes

*a. Put two electrodes into the patch to inject two pulses.*

Click the "Insert Stimulus Electrodes" button. This button will to put two electrodes into the patch, the simplest way for NEURON to inject two separately controlled current pulses. Electrode [0] (left panel) injects the first pulse and Electrode [1] (right panel) injects the second pulse.

*b. Look at the default parameters for each pulse.*

- **Delay:** The first pulse is delivered at t = 0 ms and the second at t = 10 ms.

- **Duration:** At the begin of this tutorial the default values are set so that each pulse is very short (0.1 ms); this duration is approximately that of transmitter-gated conductance changes at fast synapses such as the neuromuscular junction (explored in the Postsynaptic Potential Tutorial).
- **Amplitude:** Both of the stimulus current amplitudes are set to a large value, 125 nA. This current amplitude is approximately 1.5 times the threshold value (84.5–84.6 nA) you found for brief pulses in the Patch Threshold Tutorial (at a resting level of -65 mV).

4. **The graphs**
   Click the "Voltage vs Time Plot" and "Membrane Conductances" buttons to bring up these graphs.

# Experiments and observations

## *How long is the refractory period following an action potential?*

1. **Deliver two pulses to the patch.**
   Run the simulation. Even though the amplitudes of both stimuli are well above threshold for setting up an action potential, you should find that the second current pulse, occurring 10 ms after the first pulse, fails to generate a second spike.

2. **Measure how long the membrane is refractory.**
   Increase the delay of the second pulse until it generates an action potential. By doing this you should get a sense of how long the refractory period lasts.

3. **Determine the degree of refractoriness at various times.**
   Probe the refractoriness following the spike by decreasing the delay of the second stimulus e.g., to 10, 8, 6, and 5 ms and finding the threshold current at each of these times.

   You should find that you need more and more current—are you surprised at how much current?—as you move the second stimulus closer and closer to the peak of the action potential. No wonder an excitatory synaptic input has trouble generating an action potential when it falls in a refractory period!

4. **Question:**
   The responses to pulses at 6 and 5 ms are particularly interesting. Can you explain the waveforms of the responses at each of these times?

## *Is there a refractory period following a subthreshold depolarization such as an excitatory postsynaptic potential?*

Are the conductance increases caused by subthreshold depolarizations sufficient to have a marked effect on the excitability of the membrane?

**1. Make the first stimulus subthreshold.**

Set the amplitude of the first current stimulus to a subthreshold level such as 84 nA. Reset the amplitude and delay of the second current pulse to the default values (125 nA at 10 ms), about 50% above threshold.

**2. Run the simulation.**

Explore any refractoriness following the subthreshold event. Repeat the same exercises you did for the action potential above:

*a. Increase the time of onset of the second stimulus.*

Find when a second spike is generated. At some time point you should find that the second event lingers for quite some time at threshold. Look at the conductance changes: Are you surprised at the differences between the amplitudes of the conductances causing the subthreshold and suprathreshold events, or has the tutorial on threshold prepared you for these results?

*b. Map the refractory period.*

Determine the threshold stimulus strength required for the second pulse to generate an action potential, progressing from 10 ms to shorter onset times (e.g., 8 ms, 6 ms, and 4 ms).

The situation is perhaps more complex than you might have expected. At some time point, as the onset time of the second stimulus approaches that of the first, less current is required for the second stimulus to elicit an impulse! An <u>explanation</u> of this behavior requires that you measure the conductance changes with the crosshairs and notice their ratios at the interesting time points. Use View = Plot to expand the graph as you need to.

---

## *Are there changes in excitability following a hyperpolarizing pulse such as an inhibitory postsynaptic potential?*

**1. Generate a hyperpolarizing pulse.**

Make the first current pulse hyperpolarizing by setting its amplitude to -84 nA.

**2. Map the membrane excitability following this pulse.**

Reset the delay and amplitude of the second current pulse to the default values (10 ms, 125 nA). Use this pulse to test the excitability of the membrane following the hyperpolarizing pulse by determining the amplitude required to generate an impulse at 10 ms, 8 ms, 6 ms, and 4 ms. What do you conclude about the excitability of the membrane following a hyperpolarization?

# Patch Postsynaptic Potential Tutorial

## Synaptic transmission: Nature's way of injecting current

### NEURON's representation of a postsynaptic potential

Transmitter released from presynaptic terminals binds to postsynaptic receptors and typically causes an increase in conductance to one or more ions. The simulations in Level III will use as its model synapse the well-known conductance increase caused by acetylcholine (ACh) binding to nicotinic postsynaptic receptors at the frog's neuromuscular junction (NMJ).

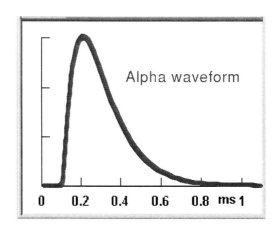

The conductance change caused by ACh has the kinetic form shown here. Since this form is often represented mathematically by a function called the alpha function, in NEURON it is referred to as the AlphaSynapse.

### Simulating postsynaptic events in a patch

This tutorial simulates the relationships between conductance, current, and voltage in a postsynaptic patch. These relationships are summarized here for reference during the tutorial.

*patch with synaptic input*

- **Postsynaptic conductance**
  The increase in conductance gated by ACh is quite brief (less than 1 ms, as seen in the figure) and has a reversal potential of -15 mV in this tutorial.
- **Postsynaptic current**
  The synaptic current flowing through the conductance of the ACh-gated channels is driven by the potential difference between the channels' reversal potential and the membrane potential; therefore its amplitude changes with changes in membrane potential.
- **Postsynaptic potential**
  In turn, this synaptic current flow causes a voltage change in the postsynaptic membrane whose time course and amplitude will depend on:
    - the time course and amplitude of the current itself,
    - the membrane properties of the postsynaptic membrane (capacitance and resistance),
    - the geometry of the postsynaptic cell.
  Although at the neuromuscular junction this voltage change is called the *end plate potential* (EPP), throughout these tutorials it will be called the *excitatory postsynaptic potential* (EPSP) as it is at neuronal synapses.

Simulating these events in a patch is formally equivalent to simulating a synapse on a soma because both the patch and the soma are isopotential. The patch may either be passive or may contain standard Hodgkin-Huxley (HH) Na and K conductances; comparing the two situations in this tutorial is instructive.

In later tutorials you will experiment with the spread of a postsynaptic potential along a dendrite (Site of Impulse Initiation Tutorial) as well as the temporal integration of more than one postsynaptic potential (Synaptic Integration Tutorial).

## Goals of this tutorial:

- To observe the relationships between synaptic conductance, synaptic current, and excitatory postsynaptic potential (EPSP)
- To explore the relationships between synaptic conductance, current, and reversal potential
- To discover whether voltage-gated channels in an isopotential region of a neuron such as a soma can distort postsynaptic potentials

## Set up to study synaptic potentials.

### 1. Assumptions

This tutorial assumes that you are familiar with all aspects of managing Neurolab simulations and are facile with arranging the windows. The Level III tutorials have a large number of panels and windows on the screen at once.

### 2. Starting the tutorial

**Start the Simulation**

Starting this simulation brings up a synaptic input panel beneath the P&G and Run Control panels, and three graphical windows described below.

### 3. The synaptic input

This menu controls the parameters of the synaptic input as follows:

- **the onset time** of the synaptic potential
- **Tpeak**, the time to peak of the synaptic conductance change
- **gmax**, the maximum synaptic conductance in μmho
- **e**, the reversal potential for the currents through this synapse, -15 mV for ACh

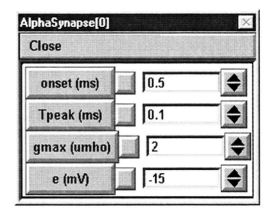

### 4. The graphs

- A Synaptic Conductance-vs-Time window shows the conductance change caused by the ACh.
- A Synaptic Current-vs-Time window plots the currents through the ACh-gated channels.
- A Voltage-vs-Time window shows the membrane potential across the patch, which is initially passive.

### 5. The Patch Parameters panel

A Patch Parameters panel does NOT appear on the monitor at first in order to keep the number of displayed panels to a minimum. This panel may be called up as needed to change or check values. The panel tells you that the patch is simulating a cell body 100 μm in length and 500 μm in diameter.

## Experiments and observations

*Observe the relation between synaptic strength (conductance) and the postsynaptic response (the EPSP).*

### 1. Stimulate the presynaptic input.

When you click R&R you will release a small amount of ACh from a pretend presynaptic terminal; in response you will see the synaptic conductance change, the resulting synaptic current, and the small EPSP. At first the patch is passive; the Hodgkin-Huxley Na and K conductances are set to zero by default. (Notice in the Reset field that under these conditions the membrane potential is -65 mV.)

### 2. Double the synaptic conductance (gmax) several times.

Double the conductance, choose Keep Lines, and rerun the simulation. Keep doubling the conductance up to 64 μmho where there is no further significant increase in the amplitude of the voltage even though the synaptic conductance continues to increase.

When Sir John ("Jack") Eccles, Bernard Katz, and Steve Kuffler observed the EPSP on frog muscle fibers <u>in 1941</u>, they had to include curare in the bathing medium to reduce its amplitude below threshold for eliciting an action potential in the muscle. You need not worry about the effects of curare because the Hodgkin-Huxley channels in your patch have zero conductance for now.

### 3. Explain the results of this experiment.

- The time courses of the synaptic currents have considerably shorter durations and different shapes from those of the conductance changes. <u>Why</u>?

- The time course of the voltage change is much slower than either that of the synaptic conductance change or of the synaptic current. Its peak is reached when the current returns to zero!

- Clearly, the voltage change approaches an asymptote as the conductance increases. What is the <u>significance</u> of this value?

**4. Compare your observations with a figure from a classic paper.**

If you rerun your simulation with Total # (ms) set to 30 ms, you can compare directly your voltage changes to those in Figure 5 of the classic study of the EPSP by Paul Fatt and Bernard Katz in 1951.

In their case, the muscle is curarized and the EPSP is small and subthreshold. If you wonder why your EPSP in a passive patch looks so similar to their EPSP in a muscle fiber with voltage-sensitive channels, you will find that experiments below and their links deal with this matter.

## Determine the reversal potential of the ACh-gated EPSP.

**1. Time the EPSP to arrive late in a prolonged current pulse.**

To observe reversal of the EPSP you will have to change the membrane potential of the patch to different values with a prolonged current pulse. The EPSP can then be timed to occur during the pulse, after it has reached a steady state.

*a. Set the Total # ms at 25 ms.*

*b. Insert an electrode into the patch.*

Bringing up the Stimulus Control panel (partially overlying the AlphaSynapse panel). Check the pulse parameters in IClamp. The duration of your depolarizing current pulse has been set at 25 ms (that is, it terminates just off the graph) and its amplitude initially is set at 5 nA.

*c. Time the EPSP.*

In the AlphaSynapse panel, change the onset of the EPSP to 20 ms so that it will occur after the current pulse has reached a steady state. Return gmax to its default value of 2 nA. Erase the traces from your previous experiment.

**2. Run the experiment.**

Deliver a set of depolarizing current pulses to the patch, increasing the amplitude of the current pulse in 5 nA steps from the initial value of 5 nA to 35 or 40 nA with each R&R. Using the "Keep Lines" option for this experiment is valuable. Where is the reversal potential?

*a. Expand the conductance and current trace amplitudes.*

The conductance and current traces will be small compared to the previous experiment. Use View = plot to expand them. What determines the amplitude of each current?

*b. Expand the time base of a selected region.*

You can expand the region around the reversal potential on any graph by choosing New View on the pop-up menu of the right mouse button and then using the left mouse button to box the interesting area. Re-size the box by pulling it out horizontally to see the events on a faster time base or vertically to have more y-axis resolution.

## *Experiment with the reversal potential of an IPSP.*

An inhibitory postsynaptic potential (IPSP) typically has a reversal potential (Erev) either at, or negative to, the value of the resting potential (Vm). Even though there is a conductance change due to the action of transmitter, there may little voltage change or even none at all if Vm is at Erev for the IPSP.

1. **Set the synaptic reversal potential "e" to the resting potential, -65 mV.**
   This is done in the AlphaSynapse menu (which may be hiding under Stimulus Control).

2. **Run the simulation with the voltage at the resting potential.**
   Do this by setting the amplitude of the current pulse to 0. Although you will observe no synaptic current and no IPSP, you will see the conductance change that will "clamp" the voltage at the value of the reversal potential and make it more difficult for depolarizing inputs to be effective. These sorts of synaptic interactions are explored in the Postsynaptic Interactions Tutorial.

3. **Reveal the IPSP.**
   An experimenter may miss observing such inhibitory inputs onto a cell. The wise investigator, however, will displace the membrane potential away from rest to test for the presence of these hidden events. Put yourself in this position and deliver a depolarizing current pulse of 5 nA, then a hyperpolarizing pulse of -5 nA to reveal the IPSP. (Remember that View = plot will allow you to view hyperpolarizing pulses that may have gone off the bottom of the graph.)

4. **Restore default parameters.**
   Return the AlphaSynapse parameters to their default values. Close the Stimulus Control panel.

## *How do voltage-sensitive channels affect the shape of an EPP or EPSP?*

The next experiments will be done on a patch containing Hodgkin-Huxley (HH) channels rather than on a passive patch. Keep in mind that the patch is equivalent to any isopotential region of a neuron such as the soma.

1. **Set the Total # ms to 10 ms.**

2. **Click the "Add HH channels" button (in the P&G manager).**
   Clicking this button will reset two parameters as follows.

   • The densities of the Na and K channels (and the leakage reversal potential) will be changed to the standard HH values.

   • The synaptic conductance, gmax, will be changed to 0.6658 µmho, below the threshold level of 0.6659 µmho for generating an impulse.

### 3. Run the simulation.

Click R&R to see the waveform of an EPSP (or EPP) in an active patch membrane. If you "increase the number of transmitter quanta" by increasing gmax the slightest bit, to 0.6659 μmho, this EPSP will generate an action potential. What a small increase this is in the conductance—only about 1 part in 7,000!

A very small difference indeed in the transmitter-gated current can make the difference between whether or not the postsynaptic membrane fires. You may remember this result from the Patch Threshold Tutorial; remember also that at threshold the decision time about whether or not to fire can be quite long, as you see here. Reflect on how many milliseconds passed between the conductance increase (and resulting current) due to the action of transmitter and the postsynaptic action potential they triggered!

### 4. Question:

Can you explain the shape of your subthreshold stimulus in this active membrane? If you are puzzled about EPP or EPSP shapes commonly observed when recording from muscle fibers or neurons (rather than a patch), a detailed explanation is available in the link. Also, consult this explanation of why EPPs have a falling phase that appears to be exponential rather than distorted like the falling phase in this active patch with HH channels.

### 5. Compare EPSPs in a passive and active (HH) patch.

If you wish to compare this EPSP in an active patch with that in a passive patch, bring up the Parameters Panel to reset the conductances of the HH channels to zero.

# Patch Postsynaptic Interactions Tutorial

In the Refractoriness Tutorial you saw that subthreshold depolarizing and hyperpolarizing currents injected through a microelectrode could influence the subsequent excitability of a neuron. In the present tutorial you will extend these observations to investigate how an excitatory or an inhibitory synaptic potential can influence whether a subsequent synaptic input causes a neuron to fire.

*patch with synaptic input*

## Goals of this tutorial:

- To see how excitatory postsynaptic potentials (EPSPs) sum in a passive membrane
- To experiment with temporal summation of EPSPs in an active soma membrane (membrane containing HH channels) in two instances:
    - when a subthreshold EPSP is followed by a suprathreshold EPSP
    - when a subthreshold EPSP is followed by another subthreshold EPSP
- To discover what can happen to membrane excitability following an inhibitory synaptic input (IPSP)

## Set up for the tutorial.

### 1. Assumptions

This tutorial assumes that you have mastered the NeuroLab manipulations and can easily keep track of many open panels and graphical windows. It also assumes that you have done the experiments in the Patch Threshold Tutorial and the Patch Postsynaptic Potential Tutorial.

### 2. Starting the tutorial

> **Start the Simulation**

Starting this simulation will bring up five panels and windows:

- two AlphaSynapse panels stacked below the P&G Manager and Run Control panels
- a Voltage-vs-Time graphical window

### 3. The two synaptic inputs

*a. Numbering of the inputs*

The two AlphaSynapse panels, [1] and [2], represent two conductance changes at the same synaptic location.

*b. Default conductances*

At the temperature of 6.3°C of this simulation, the conductance change at the threshold for generating an action potential in active membrane is between 0.504 and 0.5045 µmho. Note that the default setting of the conductance for both EPSPs is 0.5035 µmho, just below threshold.

*c. Reversal potential*

The reversal potential for EPSPs in this tutorial has been set by default to zero for simplicity.

## 4. The graphs

The Voltage-vs-Time window is initially set up for a passive patch. A similar window set up for HH channels will overlie this passive-membrane window when you click the "Add HH Channels" button in the P&G Manager. A button also exists for bringing up a Conductance-vs-Time graph.

# Experiments and observations

## *Temporal summation of EPSPs in a passive postsynaptic membrane*

### 1. Launch the Parameters panel.

The postsynaptic membrane patch may be thought of an isopotential soma 100 μm × 500 μm. Notice that this tutorial begins with both the Na and the K conductance set to 0 as appropriate for a passive membrane.

### 2. Deliver two synaptic inputs to the passive patch.

How do two EPSPs, caused by two sequential action potentials in the presynaptic neuron, interact when the postsynaptic membrane is passive? Of course there is no presynaptic neuron in this simulation: instead of shocking the presynaptic nerve you will specify the times of arrival, or onsets, of the postsynaptic potentials.

### 3. Run the simulation with the default settings.

The onset of the first EPSP occurs at 0.5 ms and that of the second at 10 ms, after the first has decayed.

### 4. Decrease the time between the EPSPs.

Gradually reduce the time of onset of the second EPSP, AlphaSynapse[2], from 10 ms to 1 ms. Is summation linear? That is, does the peak amplitude of the second EPSP increase by the amount of depolarization resulting from the the first?

### 5. Double the conductance of each EPSP.

See if your observations hold for greater conductance changes. Double, then triple the conductance for both EPSPs. Since the membrane is passive, <u>won't summation always be linear</u> for any conductance change?

### 6. Restore default settings for the next experiment.

## *Temporal summation of EPSPs in a postsynaptic membrane with HH channels: Time is of the essence!*

### 1. Add the HH channels and set the second EPSP above threshold.

Clicking the "Add HH channels" button will add the HH conductances automatically and also set the value of gmax for the second EPSP (0.555) to a value about 10% above threshold. The first EPSP remains at 0.5035, slightly below the threshold value.

**2. Open the Conductance-vs-Time and Expanded Conductances windows.**
Looking at the conductances should be useful as you consider explanations of your observations. These two graphs plot the conductances at two different gains. Both are needed since there is an order of magnitude difference in the amplitudes of the conductance changes occurring during the synaptic potential and during the action potential.

**3. Deliver the two EPSPs.**
Since the current injected that caused the first EPSP is just below threshold, the voltage lingers at threshold for some time before the K current wins over the Na current and the patch repolarizes. You should not be surprised by this long time of decision if you have done the <u>Patch Threshold Tutorial</u>. But what about the response to the second EPSP? Why does this EPSP not generate an action potential? Look at the conductances to answer this question.

(You can assure yourself that the conductance setting for this EPSP is truly above threshold by setting the gmax of the first synaptic input to zero. Remember to restore its default value afterwards.)

**4. Delay the second EPSP.**
Over what period of time does the first EPSP affect the ability of the second EPSP to generate an impulse? Generate a family of traces by delaying the second EPSP by, say, 2 ms intervals, and look at the underlying conductances.

**5. Consider mechanism and terminology.**
There are two matters to think about in this experiment. The first, of course, is <u>what mechanisms</u> can explain your observations. But the second has to do with how our terminology in neurophysiology may restrict our thinking. Wasn't the first EPSP actually inhibitory in this experiment? Perhaps a more accurate name for EPSPs would be "DPSPs," "depolarizing" rather than "excitatory" postsynaptic potentials. Since we seem to be stuck with the EPSP terminology, we hope this tutorial has at least broadened your expectations of what this event can do to the neuron.

## Combining two subthreshold EPSPs

When two just-subthreshold EPSPs occur synchronously, clearly the result will be an EPSP that generates an action potential. But what happens when the second event is delayed with respect to the first?

**1. Start with simultaneous EPSPs at 1 ms.**

Set the onset times of both EPSPs to 1 ms and restore each of their conductances to the default subthreshold value (0.5035 µmho). Plan to record both the voltage and the conductance at high and low gain for each run. Using Keep Lines is especially valuable in this experiment.

**2. Then increase the temporal separation of the EPSPs.**
Generate a family by increasing the time of onset of the second EPSP by 1 ms for each run until the action potential fails. In the low-gain Conductance-vs-Time graph (the middle graph) you see the full amplitudes of the the Na and K conductances. Looking at this graph, and at the high-gain Expanded Conductances graph, explain the changes in the amplitude of the action potential and the conductances that you observe.

**3. Further increase the EPSP separation.**
Now increase the time of onset of the second EPSP further, until at least 17 ms. First, what do you predict will happen? Are you surprised by what actually happens and can you <u>explain it</u>? Change the Total number of ms to 35 and keep increasing the separation of the EPSPs. As you watch what is happening, remember that these are two subthreshold EPSPs. To explain what you observe, measure the K conductance with the crosshairs and look carefully at the voltage at the interesting timepoints.

---

## What are the effects of an inhibitory synaptic input (IPSP) on membrane excitability?

An *inhibitory* postsynaptic potential can also affect a subsequent EPSP. It is common to find inhibitory and excitatory synapses side by side on a neuron, essentially at the same location electrically. Converting the first synapse to an IPSP here, you can experiment with its effects on a subsequent EPSP.

**1. Make the first synapse inhibitory.**
Set its onset time to the default value of 0.5 ms and set its reversal potential ("e") to minus 80 mV. You can leave its conductance unchanged at first, but experimenting with the conductance would also be instructive.

**2. Make the second event a subthreshold EPSP.**
Time it to occur at 10 ms, substantially after the IPSP. Confirm that it is subthreshold if you wish by setting the conductance of AlphaSynapse [1] to zero.

**3. Deliver the IPSP and the EPSP.**
Can you explain your observation? Hint: Before you consult the link, change the reversal potential of the IPSP and observe what these changes do to the EPSP. If the reversal potential is at the resting potential of -65 mV, for example, what is the consequence for the EPSP?

### *Voltage-dependent conductances can introduce complications into synaptic interactions.*

Your experiments should have revealed at the very least an inconsistency in the nomenclature of synaptic designations!

Activation of synapses whose reversal potentials are more positive than the resting level cause depolarizing currents to be injected into a cell, whereas synapses with more negative reversal potentials cause hyperpolarizing currents to be injected. Depolarizing currents are in the direction of eliciting a spike and have accordingly been designated as "excitatory"; hyperpolarizing currents have been designated "inhibitory."

You have seen, however, that for a period following its peak a depolarizing current can actually suppress the response of the cell to subsequent suprathreshold EPSPs under certain conditions. Conversely, an IPSP can be followed by a period of increased excitability. Nonlinear conductances are responsible.

# Passive Axon Tutorial

## How voltages spread in axons that have no voltage-sensitive channels

From the experiments in this tutorial on the passive decay of a voltage along the axon illustrated in the diagram below, you will be able to appreciate the necessity for a mechanism to enhance the voltage spread. Nature discovered voltage-sensitive ion channels and with them was able not only to enhance voltage spread but to generate a signal, the action potential, whose shape and amplitude are maintained with high integrity. This integrity will be clear in the next tutorial on propagation in a uniform unmyelinated axon (Unmyelinated Axon Tutorial).

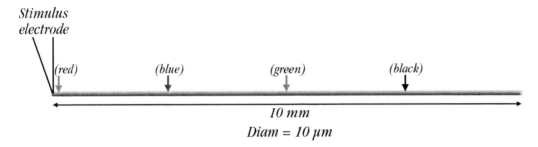

## Goals of this tutorial:

- To observe in an axon the passive spread of a voltage change in response to injected current
- To measure the "length constant" of the axon
- To experiment with how membrane resistivity and axon diameter affect the passive spread of a voltage
- To experiment with whether a change in membrane capacitance affects passive spread
- To observe passive spread when the electrode is located at different positions along this closed-ends axon

## Set up for the tutorial.

### 1. Assumptions
This tutorial assumes that you are familiar with the following panels and manipulations:

- The function of the buttons within the P&G Manager and Run Control panels
- Running simulations by clicking Reset and Run (R&R) or clicking Reset and then the "Continue for (ms)" button in the Run Control panel
- Inserting stimulating electrodes and controlling the onset, amplitude, and duration of the stimulus using the Stimulus Control panel
- Storing, erasing, resizing, and measuring values on traces in plotting windows

using the right mouse button features
- Using the field editor or the arrows box to change values in the parameters panels

If this tutorial is your introduction to Neurons in Action, we suggest that you familiarize yourself with the panels and operations listed here by clicking their links.

### 2. Starting the tutorial

**Start the Simulation**

The P&G Manager and the Run Control panel will come up.

### 3. Inserting the stimulating electrode

Press the "Stimulus Control" button (in P&G Manager) to bring up the Stimulus Control panel. The stimulating electrode (blue ball) is located at the left end of the axon and is set by default to apply a depolarizing current pulse 15 ms in duration. (You can check this value by clicking the "IClamp" diamond in the upper portion of this panel.)

### 4. Launching the first graph

Bring up the Voltage-vs-Time plot ("V vs Time, 4 locations") button in the P&G Manager. A plotting window for viewing traces of voltage versus time will appear on the upper right. You have inserted four recording electrodes into an axon 10 mm long. The locations of these recording electrodes are color-coded as follows with numbers in parentheses indicating their location:
- red: (0) (left end)
- blue: (0.25) (one-quarter of the way down the axon from the left end)
- green: (0.5 (half-way down the axon))
- black: (0.75) (three quarters of the way down the axon from the left end)

For simplicity, in this case of a passive axon, depolarizations are shown as changes from a resting level of zero.

---

# Experiments and observations

## Observe voltage responses to the current pulse.

### 1. Run the simulation.

To do this, press R&R. You will observe the voltage recordings at the four locations. How does the rate of rise and the amplitude of the voltage step change as the recording site moves farther away from the point of stimulation?

(Note that the voltage change at the recording electrode at the (0.75) position is too small to be observed at this gain.)

### 2. Increase the gain and rerun the simulation.

To increase the gain on the voltage axis, press the "V vs Time, Expanded" button. This action brings up another graph with an expanded voltage axis.

When you run the simulation again, you can note more accurately the time of onset of the voltage change as it spreads down the axon. You may be surprised to find that a voltage spreading passively has a delay associated with the attenuation of the voltage signal.

## Observe the voltage change spreading along the axon.

### 1. Call up the Voltage vs Space graph.
A unique feature of the NeuroLab tutorials is their ability to show voltages spreading or propagating in the neuron. As you prepare to watch this "movie" on this new graph, remember that the x-axis is now distance, not time. You may want to close the Voltage-vs-TIme, Expanded graph and resize the Voltage-vs-Space graph vertically for better viewing.

### 2. Press R&R to run the movie.
Displaying this movie slows down the simulation significantly because NEURON now must plot all of the voltages along the axon and then erase them before plotting them at the next integration time step. The slowness, however, should give you time to appreciate the spread of a voltage change along the passive axon and to relate these changes in space to the voltages recorded (as a function of time) at each electrode.

## Measure the length constant (L) of the axon.

### 1. What is the *length constant*?
The length constant L is not a physical property of the membrane; it is arbitrarily defined as the distance over which the voltage decays to approximately one-third of its initial value, or, more precisely, to $1/e$ (0.36788) of its initial value:
$V = (Vo) \times (\exp(-x/L)$

To measure it you will have to freeze the voltage decaying over distance.

### 2. Capture a trace from which you can measure the length constant.
Do this by stopping the movie when the voltage at the left end (red trace in the Voltage-vs-TIme plot) has reached a steady-state (maximum) level while the stimulus is still on. You will then be able to observe that the voltage in the Voltage-vs-Space plot falls off exponentially. A simple way to do this is to set the Total # (ms) button to 19; this will stop the simulation just before the end of the 20 s stimulus pulse.

### 3. Make your measurements.
- Use the <u>crosshairs</u> to measure the x and y values on the Voltage-vs-Space graph.
- Measure the initial value of the voltage at $x = 0$.
- Move the crosshairs to the voltage at $1/e$ ( 0.36788) of this initial value. The <u>spatial resolution</u> chosen for this simulation may not allow you to hit your calcu-

lated number precisely. Nevertheless, the value of L which you read nearest to this value is accurate to within 1%.

> **Caution!** *The term "length constant" is a useful characteristic for describing voltage spread in passive processes. But it applies ONLY to membranes that are linear resistances—axons with voltage-insensitive channels, or operating in a range where voltage-sensitive channels are not open. When voltage-sensitive channels are active, the current-voltage relation of the membrane is nonlinear and the "length constant" equation and name lose their meaning.*

## How does the length constant change with membrane resistance?

1. **Bring up the Axon Parameters panel.**
   The length constant depends on the ratio of the membrane resistance to the axial resistance (resistance of the axoplasm). You can change the membrane's passive resistance by changing leakage conductance in this panel to determine how such changes affect passive decay.

2. **Decrease the membrane resistance.**
   Decrease the membrane resistance by, say, a factor of 4 by multiplying its leakage conductance by this factor. Your intuition probably tells you that if the resistance is lower, current will not spread as far down the axon. But how quickly will it decline? It is instructive to actually see this experiment as a movie!

   Increase the leakage conductance four-fold by using the field editor or the UP and DOWN arrows at the right end of the value field.

3. **Run the simulation.**
   What are the new values of the initial voltage, the voltage at 1/e of the initial voltage, and the length constant? How does this value of length constant compare with your original measurement?

   If you want instead to increase the membrane resistance, the higher membrane resistance will drive your current pulse off scale. Check the link for instructions.

4. **Reset the leakage conductance to its default value.**
   Click on the red checkmark.

## How does the length constant change with axon diameter?

1. **Change the axial resistance by changing the diameter.**
   The axial resistance depends on the specific resistance, or resistivity, of the axoplasm and the diameter of the axon. Again intuition should tell you that the larger the diameter of the axon, the more easily current will flow along it, and thus the

lower will be the axial resistance. Here you can experiment with changes in diameter to get a feeling for how the voltage spread depends on diameter.

    *a. Increase the diameter four-fold to 40 μm.*

    Be aware that axons of larger diameter require more current to depolarize them to the same value as that recorded in an axon of smaller diameter (next step).

    *b. Increase the amplitude of the stimulus current.*

    We suggest an eight-fold increase to restore the voltage response to approximately the same level. (Remember: Click on the IClamp diamond in the Stimulus Control panel to gain access to the parameters of the pulse.)

**2. Run the simulation.**
What are the new values of the initial voltage, the voltage at 1/e of the initial voltage, and the length constant? How does this value of length constant compare with your original measurement? It should be <u>twice as large</u>.

**3. Reset the default values.**
Click on the red checkmarks to reset the default values of the axon's diameter and the stimulus amplitude.

## Does the length constant change with changes in membrane capacitance?

**1. Why change the capacitance?**
The capacitance of membrane is a fixed value of 1 μF/cm², so why experiment with changes in capacitance? There are two cell types in which the capacitance of a cell changes due to membrane structure. The first is the myelinated axon, which has a considerably lower capacitance due to the membrane wrappings of the myelin (capacitors in series); the second is the muscle fiber, which has a 6-fold higher capacitance because of the T-tubule system. How do changes in capacitance affect voltage spread in these cells?

**2. Change the capacitance and rerun the simulation.**
Change the capacitance by a factor of, say, 2. Does the length constant <u>change</u>?

**3. Reset the membrane capacitance to its default value.**
Click on the red checkmark.

## Move the stimulating electrode and observe the patterns of voltage spread.

Note that the ends of this axon are "sealed"— that is, they have infinite resistance. The voltage change near the end of an axon is especially affected by the resistance of the end. (This will be obvious when you move the electrode towards an end, as suggested in step #3 below).

1. **Move the electrode (blue ball) to the middle of the axon.**
   Do this in the Stimulus Control panel by clicking on the midpoint of the line representing the axon. Read the new location of the electrode just above the white lower window; it should be at the (0.5) position.

2. **Stimulate the axon at its midpoint.**
   Click R&R. Note how the voltage spread now <u>differs</u> from the pattern when the electrode was located at the left end of the axon.

3. **Move the electrode closer to one end than the other, e.g., 0.25 or 0.75.**
   Because the axon has been divided into 101 short segments for simulation, you won't be able to hit these locations precisely but you can get very close (e.g., 0.252475). Stimulate the cell at this asymmetric point.

   Observe the voltage decay away from the the point of current injection in the Voltage-vs-Space plot. The pattern of decay is quite different in the two directions.

4. **Question:**
   <u>Why</u> is there no voltage decay at the very ends of the axon (seen earliest at the end nearest the current injection)?

# Unmyelinated Axon Tutorial

In this tutorial you will begin studying the propagating action potential. The simulations here are of impulses propagating in the squid's giant axon, truly a giant at 500 µm in diameter (see diagram). By comparison, mammalian unmyelinated fibers, which carry sensations of pain and temperature, are less than 1 µm in diameter. The squid axon has provided neurobiologists with a wonderful experimental preparation. You may wonder: Why is its diameter so large?

You will see in this tutorial that rapid propagation of the impulse along an unmyelinated nerve means that the axon's diameter must be large. In mammals, most axons are wrapped with myelin to increase the speed of propagation without having to increase fiber diameter (as you will see in the Myelinated Axon Tutorial). Unmyelinated axons in mammals carry sensations that do not require action in the subsecond time frame.

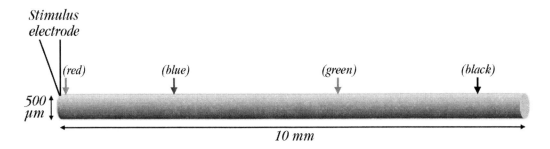

## Goals of this tutorial:

- To understand the mechanisms that underlie the shape of the propagating action potential as a function of both time and the distance along the axon
- To observe the effects of changing diameter and temperature on the shape and velocity of the propagating action potential

## Set up to stimulate the axon and view the action potential as it propagates.

1. Assumptions
   This tutorial assumes that you are familiar with the following panels and manipulations:
   - The function of the buttons within the P&G Manager and Run Control panels
   - Running simulations by clicking Reset and Run (R&R) or clicking Reset and then the "Continue for (ms)" button in the Run Control panel
   - Inserting stimulating electrodes and controlling the onset, amplitude, and duration of the stimulus using the Stimulus Control panel
   - Storing, erasing, resizing, and measuring values on traces in plotting windows using the right mouse button features
   - Using the field editor or the arrows box to change values in the parameters panels

If this tutorial is your introduction to Neurons in Action, we suggest that you familiarize yourself with the panels and operations listed here by clicking their links.

2. **Starting the tutorial**
   Click this button to start the tutorial.

   **Start the Simulation**

   - The <u>P&G Manager</u> is specialized for this tutorial with additional buttons that are specifically designed for observing impulse propagation.
   - The <u>Run Control</u> panel has an added button to allow you to change the temperature. You do not click this button; you simply change the temperature in the associated field.
   - A <u>Stimulus Control</u> panel and a Voltage-vs-Time graph also come up and are described below.

3. **The stimulating electrode**
   The "Location" of the stimulating electrode (blue ball) is represented in the Stimulus Control panel as the fraction of the distance along the axon from "0" at the left end to "1" at the right end (as it was in the Passive Axon Tutorial). As shown in the diagram below, the electrode is located at the extreme left end of the axon (represented below the "Location" button as "axon(0)").

   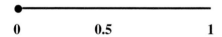

   **0                0.5                1**

4. **The first graph**
   A <u>plotting window</u> for viewing a single Voltage-vs-Time trace comes up automatically; a recording electrode has been inserted into the center (the 0.5 position) of the axon. (This location of the recording site is printed on the graph as "axon.v(0.5)".) During the tutorial this graph will be overlaid with one in which two electrodes are inserted, and finally with one in which four electrodes are inserted (as shown in the diagram above).

---

## Experiments and observations

*Record the action potential as a function of time at various locations along the axon.*

1. **Record the action potential at one location.**
   Press R&R. The axon is stimulated at its left end and the action potential is recorded at the center of the axon. Would you see a similar recording if you stimulated the axon at its right end? Do the experiment: Move the electrode by clicking on the line at the right end. (Don't forget to move it back afterward).

**2. Record the action potential at two locations.**
Press the "Voltage vs Time, Dual traces" button; the graph you call up will overlie the previous graph. Two recording electrodes are now inserted, located near the two ends of the axon (location 0.1 and 0.9). Since the stimulating electrode is located at the left end, the action potential will arrive earlier at the red recording electrode than at the black one. Run the simulation. Move the stimulating electrode to the right end of the axon and run the simulation again. Move it back for the next experiment.

**3. Record the action potential at four locations.**
Press the "Voltage vs Time, Quad traces" button, then R&R, to do this experiment with four recording electrodes spaced along the axon at positions 0.01, 0.3, 0.6, and 0.9. From these plots see if you can envision the membrane potential distribution as a function of space. This mental task is not easy. Here is where the computer can really show off, as you will see below.

**4. Observe the currents and conductances underlying the propagating action potential.**
Close the Dual and Quad trace graphs (leaving the graph in which the membrane potential is recorded at the center of the axon). Press the "Membrane Currents" and "Membrane Conductances" buttons to bring up graphs in which you can view the Na and K currents and conductances at the center of the axon. Run the simulation and look carefully at the temporal relations of the voltage, currents, and conductances.

**5. Questions:**
Do the ionic current patterns for this propagating action potential appear similar to those for an action potential in a uniform patch, which you observed in the Patch Action Potential Tutorial? You will see below that the advancing impulse propagates because of the longitudinal currents that flow from itself to depolarize the membrane ahead. How can it be, then, that the currents in a propagating impulse at any given point in the axon have time courses similar to those for a stationary impulse in a patch?

**6. Prepare to see a movie of the impulse propagating.**
Close the Membrane Currents and Membrane Conductances windows. Leave open the Membrane Voltage-vs-Time (Single Trace) plot. (If you accidently close this plot, you must quit and then re-start the simulation to get it back.)

*Display the impulse as it travels along the axon (voltage as a function of space).*

**1. Bring up the Voltage-vs-Space graph.**
Notice that the axon is 10,000 μm (10 mm) long; the four locations at which the voltage is recorded as a function of time are indicated as colored arrows along the x-axis. If you reopen the Voltage-vs-TIme, Quad Traces graph you will observe traces that are color-coded to these arrows.

**2. Press R&R to run the movie.**
You will generate an action potential with a depolarizing current pulse at the left end of the axon and see the impulse propagate from left to right. You may be surprised by what you see: the voltage distribution is almost uniform along the axon! (As explained in the <u>movie section of the Passive Axon Tutorial</u>, displaying the movie slows down the simulation significantly compared to simply recording the impulse as a function of time.)

**3. Relate what you see in the movie to the Voltage-vs-Time plots.**
- Stop the action potential about half-way through its travel and be certain you know where its rising phase is on the Voltage-vs-Space plot as well as on the Voltage-vs-Time plots.
- Try to follow the voltage at one spatial point as it moves up and down with time.

**4. Follow the action potential in a step-pause manner.**
- First press Reset (not Reset and Run) in Run Control.
- Then press the "Continue for (ms)" button to advance time in short increments, for example 0.1 ms. Each time you press this button the movie will advance another 0.1 ms and pause.

---

## Observe the effect of changing the axon diameter on impulse propagation.

**1. Select Keep Lines in the Membrane Voltage-vs-Time (single trace) graph.**
With the right mouse button and the cursor on the graph, select the Keep Lines feature so that you can compare the shapes of action potentials generated with different axon diameters.

**2. Open the Axon Parameters panel to change the diameter.**
By how much must you reduce the diameter of the axon so that you can see more of its waveform in the movie? When you reduce the diameter, why can you see more of its waveform?

As the axon diameter is decreased, less current is necessary to stimulate it. (<u>Why is this</u>?) The current, then, may be reduced to minimize the current "shock artifact" during the stimulus by selecting IClamp in the Stimulus Control panel and reducing the amplitude of the current pulse from 20,000 nA to a lower value .

**3. Question:**
In the Voltage-vs-Time graph, notice what happens to the delay of the action potential with respect to the stimulus and also whether the shape of the action potential is affected when the axon's diameter is changed over two log units. Can you <u>explain</u> your observations?

**4. Prepare for the next experiments.**
Restore the default diameter and erase the Voltage-vs-TIme traces. Continue to Keep Lines.

## *Observe the effect of changes in temperature on the propagation of the impulse.*

1. **Warm the axon by several increments of 5°C each.**
   Store the Voltage-vs-Time traces with Keep Lines. How is the action potential affected? Click here for a discussion.

2. **Continue to increase the temperature in 5°C steps.**
   As you increase the temperature, do you find a point where the action potential fails? Exactly how does it fail, and why? Click here for the answer.

3. **Restore the default temperature to prepare for the next experiment.**

## *Measure the velocity of propagation of the impulse.*

1. **How does one measure velocity experimentally?**
   Two electrodes are inserted into the axon a known distance apart and the time at which the impulse crosses some chosen voltage is measured at each location. The velocity may then be calculated since velocity is equal to the distance between the electrodes divided by the time that it takes for the impulse to travel from one electrode to the next (mm/sec). You can do this with the Voltage-vs-Time, Dual Traces graph, using the crosshairs to measure the time at which each action potential crosses zero.

2. **Call up the Voltage-vs-Time, Dual Traces graph and measure velocity.**
   Using the difference in the time of zero-crossing for the impulse at each of these recording sites, and the 8,000 μm length between them, calculate the impulse conduction velocity (V = length/time). You will have to put up with a certain imprecision due to the time resolution of the points plotting the action potentials.

3. **Measure the velocity as a function of axon diameter.**
   For small diameters you may notice that the action potential is so slow that it does not pass the 0.9 position in the 5 ms allotted on the plot. You can set the Total # ms (in Run Control) to a greater value (e.g., 10 ms). And remember to decrease the amplitude of the stimulus current as you decrease diameter.

   Keep your plot for comparison with a similar plot you will make in the next section for myelinated nerve. There is a critical diameter below which myelination confers no advantage. You will be able to determine this value.

# Myelinated Axon Tutorial

The following simulations are of high-speed impulse propagation in a single, myelinated axon of the frog, diagrammed below.

There are 10 myelinated regions (internodes), each 1000 μm (1 mm) long, and 11 nodes, each 3.2 μm long. Thus the axon's total length is 10,035 μm, marginally longer than that of the squid axon in the Unmyelined Axon Tutorial. Note that the nodes make an insignificant contribution to the axon's total length.

This axon's diameter is a typical 10 μm rather than the squid's extraordinary 500 μm. In spite of the 50-fold reduction in its diameter, the many wraps of the membrane making up the myelin boost the frog fiber's velocity to rival that of the squid!

Myelination alters the axonal distribution of channels. Within the myelinated regions, the normal Hodgkin-Huxley ion-channel densities have been assigned. At the nodes the densities are 10-fold greater than normal, matching experimental observations and simulation analysis.

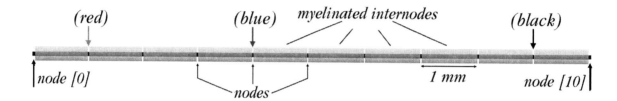

## Goals of this tutorial:

- To understand the shape of the action potential propagating in myelinated nerve as a function of time and of distance along the axon
- To measure the velocity of impulse propagation in myelinated nerve and compare it to unmyelinated nerve
- To change the degree of myelination and the temperature, two clinically important factors, and observe the effect on conduction velocity

## Set up for the tutorial.

### 1. Assumptions
This tutorial assumes that you are now familiar with the manipulations of the NeuroLab tutorials, summarized in point #1 (Assumptions) of the setup instructions for the Patch Action Potential Tutorial. Consult the links for help when needed.

### 2. Starting the tutorial

### Start the Simulation

A P&G Manager with specific buttons for this tutorial comes up along with the Run Control and Stimulus Control panels.

3. **The stimulating electrode**
   The stimulating electrode is located in the center (0.5) of node[0], the node at the extreme left end of the axon (see diagram above). This location is specified as node[0](0.5) just above the lower portion of the Stimulus Control panel.

4. **The graphs**
   Launch the Voltage-vs-Time and Voltage-vs-Space graphs when called for in the tutorial steps. The color-coded, voltage-recording electrodes are located as shown in the diagram above.

## Experiments and observations

### *Observe the shape of the impulse in a myelinated axon.*

1. **Record the action potential at Node[4].**
   Call up the Voltage vs Time, Single Trace graph. When you run the simulation, you will record the action potential in the middle (0.5) of node[4].

2. **Press Node Currents (P&G Manager) to view INa and IK at Node[4].**
   Is there any marked difference in shape, timing, or amplitude between the currents flowing into the node and the currents you observed to underlie the action potential in the Unmyelinated Axon Tutorial? Should there be?

### *Observe impulse propagation in a myelinated axon.*

1. **Call up the "Voltage-vs-Space" graph and run the simulation.**
   The voltage distribution will propagate along the axon from the stimulating electrode at left to the right. Are you surprised by its peculiar ratchet-like appearance? Use the "Stop" button to freeze the action potential and examine these bumps more closely. Are they real or an artifact of doing experiments with a simulator?

   Remember that you can also pause the movie at selected intervals by clicking Reset, then the "Continue for (ms)" button, typing the interval in the field of this button.

2. **Record the impulse at Nodes [1], [4], and [9].**
   Call up the Voltage-vs-Time, Dual Traces graph. It will overlie the Node Currents graph. Recording electrodes are now located at Node[1] and Node [9] in addition to the one in the uppermost graph at Node[4]. When you run the simulation, make certain the time at which each electrode records the impulse makes sense.

### Measure conduction velocity and determine how the degree of myelination affects velocity.

1. **Use the recording electrodes at Nodes[1] and [9] to measure velocity.**
   As you did in the Unmyelinated Axon Tutorial, measure with the crosshairs the time at which each action potential reaches some reference point; in this tutorial it is best to choose the peak. The recording electrodes are approximately 8,000 μm apart. Using this information, calculate the velocity.

2. **What is the relation between number of myelin wraps and the axon's capacitance and conductance?**
   You can change the number of wraps of myelin in the field associated with that button in the P&G Manager. But it is interesting as you change the number of wraps to follow how wrapping affects the membrane's net capacitance and conductance. You can observe these parameters in the Myelinated Region Parameters panel. Start with zero wraps and then increase the number, say, by ones or tens.

   When you set the number of myelin wraps to "0", you of course have an unmyelinated axon so the membrane has a capacitance (Cm) of 1 μF/cm$^2$. As you add wraps, the capacitance is reduced to 0.5 μF/cm$^2$ with one wrap, again half of that with two wraps, etc., since the net capacitance of capacitors in series is the sum of their reciprocals. Since the wrapping membrane also has resistance, and the net resistance of resistances in series is the sum of their individual resistances, the effective resistance is increased in proportion to the number of wraps. Correspondingly, the membrane's conductance (the reciprocal of resistance) decreases with wrapping.

3. **Now determine the velocity of the impulse as a function of the number of wraps.**
   Measure and plot (manually) the velocity versus the number of wraps. You will need to adjust the stimulus amplitude as the number of wraps is decreased. This experiment should give you an appreciation of the effectiveness of myelin. Does increasing the number of wraps above 150 offer a proportional increase in the velocity of propagation?

4. **Restore the number of wraps to the default value of 150.**

### What is the effect of temperature on the propagation velocity?

1. **Compare the velocity in myelinated frog axon with that in unmyelinated squid axon.**
   Do the small, myelinated axons of frogs conduct at roughly the same velocity as the unmyelinated huge axons of squids? To compare the velocity of myelinated frog axon to the velocity you calculated for the squid axon, cool the myelinated axon to 6.3°C (the temperature for squid axon) and determine the velocity.

By how much is the impulse in myelinated axon slowed when it is cooled from the default temperature of 22°C to 6.3°C?

2. **Plot the velocity as a function of temperature.**
   Cool and warm the myelinated axon by increments or decrements of 5 or 10°C. This experiment assumes that you have already experimented with the effects of changing temperature on the action potential in the Patch Action Potential Tutorial and the Unmyelinated Axon Tutorial. If you have not, and you wish to know more about the effect of temperature on the action potential, consult this link.

# Partial Demyelination Tutorial

This tutorial simulates an action potential in myelinated nerve attempting to propagate through a demyelinated (bare) region, as in multiple sclerosis (MS).

The axon is 10,000 μm long as shown in the diagram. The right half is myelinated, with 5 node/myelinated internode pairs; the left half is demyelinated. In the tutorial you will first insert a stimulating electrode into the left end of the bare axon, then move it to the right end of the myelinated region (node[4]).

Ion channel densities have been assigned on the basis of experimental observations. The densities in the nodes are 10-fold greater than those in squid axon. The bare axon has been assigned the normal density of channels in unmyelinated nerve, distributed uniformly. The distribution of channels in demyelinated axon is only <u>partially understood</u> at present and depends on how long the axon has been demyelinated.

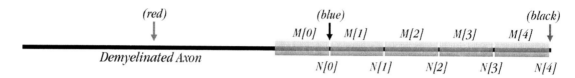

## Goals of this tutorial:

- To observe the shape of the action potential as it travels from the unmyelinated region of the axon into the myelinated region
- To observe the features of the action potential as it tries to invade the bare axon from the myelinated region
- To observe how changes in the ion conductances in the bare axon and also in the temperature affect the ability of the action potential to invade the bare axon from the myelinated portion

## Set up for the tutorial.

### 1. Assumptions

This tutorial assumes that you are familiar with the following panels and manipulations:

- The function of the buttons within the <u>P&G Manager</u> and <u>Run Control</u> panels
- Running simulations by clicking Reset and Run (R&R) or clicking Reset and then the "Continue for (ms)" button in the Run Control panel
- Inserting stimulating electrodes and controlling the onset, amplitude, and duration of the stimulus using the <u>Stimulus Control</u> panel
- Storing, erasing, resizing, deleting, and measuring values on traces in <u>plotting windows</u> using the <u>right mouse button</u> features
- Using the <u>field editor or the arrows box</u> to change values in the <u>parameters panels</u>

If this tutorial is your introduction to Neurons in Action, we suggest that you familiarize yourself with the panels and operations listed here by clicking their links.

## 2. Starting the tutorial
Click this button.

> ## Start the Simulation

A P&G Manager specific for this tutorial and the Run Control panel will come up. You will click buttons in the P&G Manager to bring up other panels.

## 3. The stimulating electrode
Two Stimulus Control buttons are available in the P&G Manager, one to put the stimulating electrode (Stimulus 'Trode) in the left end of the bare axon and the other to put it in the right end (node[4]) of the myelinated segment. Clicking either of these buttons will bring up its own Stimulus Control panel.

Although the electrode may be moved after insertion as usual by clicking on the line representing the axon in this panel, it is preferable to switch locations using these buttons because the amplitude of the current stimulus needs to be different in the bare axon and in the myelinated segment.

## 4. The graphs
Press the "Voltage-vs-Time" and "Voltage-vs-Space" buttons to call up these two graphs. In the Voltage-vs-Time graph, the traces are color-coded to the three recording sites shown in the diagram above and on the axis of the Voltage-vs-Space graph:
- the center of the bare half of the axon
- at node[0]
- at node[4]

---

# Experiments and observations

*Observe impulses traveling in partially demyelinated axons.*

### 1. Stimulate the left end of the demyelinated (bare) axon.

*a. Press the "Stimulus 'Trode in Bare Axon" button (in the P&G Manager).*

The Stimulus Control panel that comes up shows a stimulating electrode in the left end of the bare half of the axon. The amplitude of the current pulse has been set to a large enough value to evoke an action potential in the bare portion.

*b. Run the simulation (press R&R).*

The impulse will be initiated at the far-left end of the demyelinated region and travel to the right into the myelinated region.

In the Voltage-vs-Time graph, notice that the second and third recorded impulses (recorded at the nodes) are much closer together than are the first and second recorded impulses (recorded in the bare area and at the first node). Why is this?

*c. Relate the Voltage-vs-Time recordings to the Voltage-vs-Space movie.*

As you did in the <u>Unmyelinated Axon Tutorial</u>, watch the voltage at a point in the Voltage-vs-Space graph as it moves up and down in time. If you pick a point where a recording electrode is located (for example, in the middle of the bare axon), you can compare it to the trace in the Voltage-vs-Time graph.

To slow down the process, click the "Continue For (ms)" button in the Run Control panel to play and pause the movie repeatedly. Or click the "Stop" button to stop the movie at any time.

*d. Stop the impulse in the bare axon.*

Make certain you understand the shape of the action potential in both plots. For example, which phase is the rising phase in the Voltage-vs-Time plot and in the Voltage-vs-Space plot? Can you explain the <u>shape of the rising phase</u> in the Voltage-vs-Space plot? If you understand the shape, you probably have a good understanding of how the impulse propagates.

*e. Close the Stimulus 'Trode in Bare Axon panel.*

This action will remove the stimulating electrode. You will next insert it at the other end of the axon. Although you can move the electrode to the other end of the axon by clicking on the line, the amplitude of the stimulus will then be inappropriately large for stimulating the myelinated segment, as mentioned above.

 **Attention!** *Next you will put the stimulating electrode in node[4]. It is necessary to close the Stimulus 'Trode In Bare Axon panel to avoid generating impulses at both ends.*

2. **Reverse the direction of stimulation: Excite the myelinated region.**
Press Stimulus 'Trode in Node[4] to insert the stimulating electrode at the rightmost node of the myelinated region. Run the simulation. What happens to the impulse at the junction between the myelinated and bare axon? <u>Explain your observations</u>. Relate what you see in the Voltage-vs-Space movie to your recordings with the three electrodes in the Voltage-vs-Time graph.

---

## A change in temperature is known to improve the condition of multiple sclerosis patients. Investigate the basis for this phenomenon.

1. **Make an educated guess.**
Would you expect warming or cooling to improve the prospects of impulse invasion of the demyelinated region? Recall your experiments on the effect of temperature on the <u>patch action potential</u>.

Hint: What change will enable the action potential in the myelinated segment to supply more current to the bare axon?

2. **Test your hypothesis.**
Change the temperature (in the Run Control panel). Run the simulation to see if impulse invasion of the demyelinated region improves or worsens. A detailed discussion of the connection between <u>temperature, threshold, and impulse propagation</u> is available.

(Although the temperature range in which you are experimenting is appropriate for frog axon and not for humans, the principle is the same for both species.)

3. **Question:**
What is the smallest change in temperature required to produce any difference you observe? Your observations have a clinical correlation in the <u>Uhthoff phenomenon</u>.

4. **Observe impulse resurgence in the myelinated axon.**
Note that at a certain critical temperature the conditions are just right for the impulse in the myelinated axon to <u>resurge in the myelinated region</u>; in the Voltage-vs-Time graph you should be able to see two action potential peaks at node[4] (blue trace).

5. **Restore the temperature to the default value of 25.2°C.**

## What changes in axon parameters will permit the impulse to invade the bare region?

1. **Launch the Bare Axon Parameters panel.**
In this panel you can alter parameters of the demyelinated portion of the axon. When you have changed a parameter, remember to reset it before changing another one. You can experiment with changes that will promote impulse invasion of the bare axon.

a. *Change the density of functional K channels.*

For a hint read a <u>quote</u> from the National Institutes of Health web page entitled "Therapy to improve nerve impulse conduction."

b. *Change the density of functional Na channels.*

By how much must you change the density to enable invasion of the bare axon?

c. *Change the diameter of the bare axon.*

In what direction would you expect a change to facilitate impulse invasion?

Hint: Any change that increases the longitudinal current supplied to the bare region of axon from the myelinated region will assist the struggling action potential to become regenerative.

*d. Prepare for the next experiments.*

Be certain to restore all bare axon parameters to their default values. You can close the Bare Axon Parameters panel or leave it open.

## 2. Change parameters of the myelinated axon.

Click the "Internode Parameters" button. Four menus will come up in a "tray" for four of the five internodes (M[0] through M[3] on the diagram above). The far right internode, M(4), is left out. You can adjust the length of each internode, its degree of myelination (the capacitance, which is 1 µF/cm$^2$ divided by the number of wraps), and the inside diameter of the axon (the diameter of the axon without its wrapping).

*a. Experiment with the internode M[O].*

What change in the parameters of this adjacent internode will increase the longitudinal current into the bare axon and cause the action potential to propagate there?

- Questions: What if you <u>change the length of this myelinated segment</u>? Should it be longer or shorter to supply more current to the bare axon? How much change is needed?

- Question: What if you <u>change the degree of myelination</u> of this segment?

- Question: Will <u>changing the diameter</u> of this one segment have an effect?

*b. Change parameters of the other internodes.*

How crucial is the adjacent internode compared to the more remote internodes? Experiment in a similar fashion with the other three internodes.

# Axon Diameter Change Tutorial

What happens to the action potential when it encounters a branch point or when it tries to invade a cell body? This tutorial explores the transmission of impulses through a region of a cell where there is an abrupt change in diameter (a 10-fold difference in the default example). Recordings are made at three locations in the smaller axon and two locations in the larger axon (see diagram).

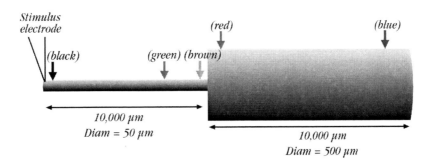

The smaller-diameter (50 μm) segment on the left may be thought of as a branch of the larger-diameter (500 μm) axon on the right. Indeed, the 500-μm-diameter squid giant axon often has 50-μm-diameter branches.

The inhomogeneous axon in this tutorial may also be thought of as an axon and cell soma.

## Goals of this tutorial:

- To observe impulses in the smaller axon struggling to invade the larger axon and understand the curious current patterns that result from the struggle
- To explore how the ratio of the two diameters affects the ability of the action potential to invade the larger-diameter axon
- To experiment with how temperature changes affect invasion of the larger-diameter axon
- To understand why impulses move with no difficulty from the larger to the smaller axon

## Set up for the tutorial.

### 1. Assumptions
This tutorial assumes that you are completely familiar with the manipulations of the NeuroLab tutorials. A familiarity with the Partial Demyelination Tutorial would also be helpful.

### 2. Starting the tutorial

**Start the Simulation**

### 3. The stimulating electrode

Bring up the Stimulus Control panel. In the present tutorial, the stimulating electrode is at the left end (0) of the small axon (red portion of line). The line is red to indicate that the stimulating electrode is inserted into that component. The black line represents the large axon until you move the electrode to this segment later in the tutorial, when that portion will turn red.

### 4. The graphs

a. *Bring up the Voltage-vs-Time and Na Current-vs-Time graphs.*

With five recording sites (color-coded as in the diagram above), you will have to work harder than usual to keep all of the traces straight. Remember that you can delete traces by choosing Delete with the right mouse button and clicking on their labels, shown in spatial order at the right margin of each graph (leftmost recording site at the top). To restore the traces, however, you must close and reopen the graph.

b. *Bring up the Voltage-vs-Space graph.*

The sites of the recording electrodes are indicated by arrowheads along the x-axis.

---

## Experiments and observations

*Observe propagation from a smaller diameter process into a larger diameter region.*

### 1. Observe the impulse in the Voltage-vs-Space window.

Run the simulation. There are many things to notice here as the impulse moves into the large-diameter axon and then is initiated again in the smaller axon

- the large stimulus artifact at the left end of the small axon setting up the spike
- the change in steepness of the voltage distribution as the impulse encounters the junction
- the voltage profile in the large axon as the voltage hesitates at threshold
- the profile as it finally gives an action potential there
- the changing amplitude at the junction as it establishes itself again in the small-diameter axon

To understand these changes it will be useful to look at the graphs of voltage and current versus time.

### 2. Study the traces in the Voltage-vs-Time graph.

It probably is most useful to run the simulation over and over and watch each trace as the action potential moves from left to right and then bounces back from right to left. Of course the black electrode will record the voltage first—and last. Relate the voltage recordings to the movie.

- Question: Is it clear from the movie why the second green action potential has a smaller amplitude than the first?
- Question: Is it clear why the red recording electrode begins to record a depolarization before the blue electrode but then records the action potential later than the blue electrode?

### 3. Study the Na currents.
You should recognize the "kinky" shape of the Na current from the Action Potential Tutorial.

- Question: What is the most eye-catching current in this group of current recordings?
- Question: Why is the Na current larger near the junction?

## How does propagation from the smaller into the larger axon depend on the ratio of the diameters?

### 1. Bring up the Small Axon Parameters panel.
Experiment with diameter to see how the diameter of the small axon affects the ability of the impulse to propagate into the large axon. You may want to delete all of the current traces except the interesting ones (brown and red) at the junction and then use Keep Lines to compare simulations at different diameters.

(Remember that you delete traces by choosing Delete with the right mouse button and clicking on the label; you restore them by closing and reopening the graph.)

### 2. Reduce the diameter.
The default diameter ratio was set near the critical value. How small a reduction in the small axon's diameter will cause failure of invasion of the large axon?

### 3. Increase the diameter.
What magnitude increase will assist propagation into the large axon? Must the diameters of the two processes be much closer in size to one another or will merely a small increase in the diameter of the small axon be sufficient?

### 4. Experiment with other variables.
We have left other parameters in this menu for you to change if you wish. A menu for changing parameters in the large-diameter axon may also be called up. What happens to propagation if you change the capacitance of either of these segments? What happens if you change a channel density? Using impulse invasion as an assay, you can thoroughly test your knowledge of neurophysiology!

## *Investigate how invasion depends on temperature.*

### 1. Restore default parameters.
Reset the diameter of the small axon to the default value of 50 µm.

### 2. Think, then change the temperature.
Will cooling or warming improve transmission from the smaller to the larger axon? Recall your experiments with temperature in the <u>Patch Action Potential Tutorial</u> and the <u>Partial Demyelination Tutorial</u>. Will the same principles apply to the current experiment? A <u>discussion</u> of the effects on temperature on the impulse and its propagation is available.

### 3. Study the voltage change and the Na current at the junction.
If you haven't done so by now, delete all of the traces in the Voltage-vs-Time and the Current-vs-Time plots except for the brown traces recorded in the small-diameter axon at the junction. Use Keep Lines to compare the recordings at several temperatures. How sharp is the temperature dependence of impulse invasion?

## *Reverse the direction of propagation.*

### 1. Move the electrode to the far right end of the large axon.
If you now run the simulation, you will find that there is insufficient default-stimulus current to generate a spike in the large axon. <u>Why</u>?

### 2. Increase the stimulus current amplitude 10-fold.
You will see that the impulse is generated and travels smoothly from right to left with no hesitation; the only noticeable change is that the wavelength of the spike (or spatial extent) becomes shorter in the small-diameter region. (<u>Why</u> is this the case?) The Na current density is almost constant in all locations. In other words, in this direction, the impulse traverses the diameter change with ease. If you have worked through the exercises of this tutorial this observation should be completely understandable.

# Non-uniform Channel Density Tutorial

Can an action potential propagate through a region whose channels are blocked by anesthetics or toxins? In this tutorial you will explore how a change in the density of functioning channels affects impulse propagation through the region containing those channels.

This tutorial consists of three squid axon segments, each 10,000 μm (10 mm) long. Records of the membrane potential are taken in the centers of each of the segments as indicated by the arrows on the diagram. Remember that NEURON views channel density and conductance as equivalent: Blocking a fraction of the channels, for example with an anesthetic, is the same as decreasing the density of functioning channels as shown on the diagram.

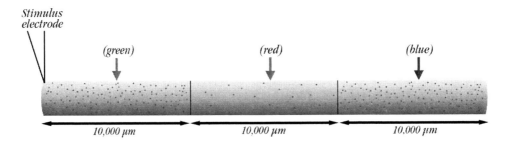

## Goals of this tutorial:

- To test how local application of tetrodotoxin, a Na channel blocker, affects propagation
- To test how local application of 4-aminopyridine, a K channel blocker, affects propagation
- To test how application of the anesthetic lidocaine, which affects both conductances, alters impulse propagation through the anesthetized region
- To test how local trauma affects propagation

## Set up for the tutorial.

### 1. Assumptions
This tutorial assumes that you are familiar with all of the manipulations of the NeuroLab tutorials.

### 2. Starting the tutorial

**Start the Simulation**

### 3. The stimulating electrode
Launching the Stimulus Control panel shows that the electrode is located at the left end of the axon. The leftmost third of the line is red, indicating that this is the segment under control of the stimulus. Check the pulse parameters: You will be stimulating with a short, sharp shock.

## 4. The graphs

*a. Bring up the Voltage-vs-Time and Na-Current-vs-Time graphs.*

The color-coded labels of the three recording electrodes are stacked in spatial order, with the leftmost electrode at the top. Run the simulation to see the three recordings of the action potential. Your trained neurophysiological eye will probably notice immediately that the peak amplitude of the action potential increases and the peak Na current decreases ever so slightly as the impulse travels from left to right (you can check this out with the crosshairs). This results from the way the tutorial was set up—can you <u>guess why</u>?

*b. Bring up the Voltage-vs-Space graph.*

Arrowheads color-coded to the voltage-vs-time traces indicate the position of the recording electrodes.

# Experiments and observations

## *Block Na channels with tetrodotoxin (TTX).*

### 1. Apply a toxin highly specific for Na channels to the middle segment.
Bring up the Middle Axon Parameters menu. Use <u>tetrodotoxin (TTX)</u>, a very powerful toxin made by the puffer fish (about 1 mg is a lethal dose for a person). Reduce the maximum Na conductance several fold in the middle axon segment.

### 2. How much block can the axon tolerate?
Observe carefully the voltage patterns as the impulse in the middle segment begins to fail. Find out by how much you can block the Na conductance and still be able to recover an impulse in the right axon segment. Use Keep Lines to compare voltage and current traces as you gradually block the Na conductance.

You may have found a particular value for the Na conductance that caused a reflection from the right segment back into the middle segment. Why does this happen? Hint: Note the timing of the action potential in the rightmost segment in the normal and blocked conditions.

### 3. Restore default values.
Remember to restore the default value of the conductance(s) after this and each of the following experiments.

## *Block K channels with an aminopyridine.*

### 1. Apply a K channel blocker to the middle segment.
Apply 4-aminopyridine (4AP), a blocker specific for the Hodgkin-Huxley K channel. Reduce the K conductance in the middle section.

### 2. <u>Interpret your observations.</u>

## *Block both Na and K channels with local anesthetics.*

### 1. Apply lidocaine to the middle segment.
Anaesthetics such as procaine and lidocaine affect both the Na and the K conductances about equally. Reduce both the Na and K conductances 2-fold, 5-fold, and 10-fold. Guess what will happen before running the simulation in each case—the result may not be as straightforward as you think it will be.
  • Question: Can you explain <u>why the impulse is delayed</u> in propagating into the right-hand section?

### 2. Compare blocking both conductances together with blocking them separately.
  • Question: Can you explain the <u>effects of lidocaine</u> on the amplitude and particularly on the shape of the impulse?

## *Investigate the effects of trauma.*

Mechanical trauma can damage nerve membranes and their proteins, making them more leaky to ions and overwhelming their pumps, which maintain the intracellular [Na] at a low value and [K] at a high value.

### 1. Damage the middle segment.
Simulate an increased leakiness (gl ) by 2-fold, 5-fold, 10-fold, and 50-fold. By how much can you increase the leakage conductance and still maintain transmission to the right segment?

This experiment should give you an idea of how much damage an axon impaled with a sharp microelectrode can sustain if the electrode causes damage in the process of insertion. Notice how much damage is necessary before you see depolarization at the site of damage.

### 2. Block the Na pump in the middle segment.
Simulate damage to the Na pump in the middle segment by reducing ENa by e.g., 5, 10, 20, and 50 mV. When does transmission fail?

### 3. How much length can be damaged before transmission fails?
With ENa reduced, increase the length of the middle segment. When you do this, the left segment will be forced off the graph to the left on the Voltage-vs-Space plot and you will have to wait for the action potential to arrive on the plot after the stimulus pulse is given. If you have made the middle segment very long, you will have to increase the time of running the simulation in the "Total # ms" field (RunControl). Are you surprised to learn how much local damage a nerve can tolerate?

The View = Plot option will allow you to see the entire length of the axon but the labels for the recording electrodes will no longer be at the correct location.

## Conclusions

This tutorial reveals that there is a lot of resilience built into the impulse-generating mechanism. A large "safety factor" ensures transmission in the face of considerable difficulties. In every case in which the right segment is activated, the shape of the action potential there is quickly restored back to that in the initiating segment.

## Play with NeuroLab.

### 1. Change parameters of the end segments.
Click on the "Left & Right Axon Parameters" button to bring up panels in which to change the values of parameters in these segments. They afford the possibility of exploring a large variety of interactions.

### 2. Change the temperature.
Investigate the effect of a temperature change combined with local changes in conductance. Your experience, gained in earlier tutorials, ought to allow you to predict the outcome of such experiments.

# Site of Impulse Initiation Tutorial

This is the first tutorial in the final section of *Neurons in Action* that deals with how action potentials initiate and propagate in neurons. In this tutorial you can see how the geometry of a neuron affects its site of impulse initiation.

Consult <u>CELLS</u> for an introduction to this topic.

## Geometry of the neuron in this tutorial

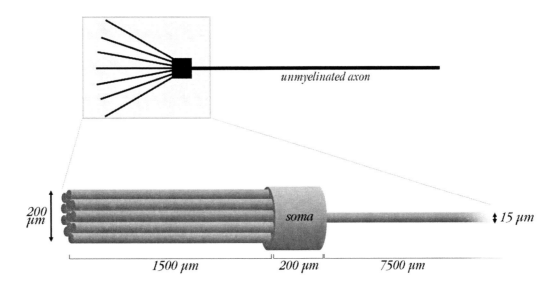

In the following experiments you will simulate voltage changes in a neuron with the simplest possible morphology comprising the following regions (from left to right on the diagram):

- a single, passive dendrite representing the entire dendritic arbor of the neuron <u>collapsed</u> into a cylinder as shown in the diagram
- a soma with Hodgkin-Huxley channels
- an *unmyelinated* axon with Hodgkin-Huxley channels
- a single synaptic input onto the dendrite (not shown on the diagram)

## Goals of the tutorial:

- To interpret movies of impulse initiation and voltage distribution along the cell in response to synaptic input onto its dendrite
- To experiment with the effects of the following parameters on the voltage distribution:
  - the strength of the synaptic input on the dendrite
  - the location of the synaptic input on the dendrite
  - changes in the diameter or length of the soma
  - changes in the diameter of the axon
  - changes in ion channel densities

# Set up for the tutorial.

### 1. Assumptions
This tutorial assumes that you are now a whiz with NeuroLab manipulations. It also assumes that you have worked through the <u>Patch Postsynaptic Potential Tutorial</u>.

### 2. Starting the tutorial

## Start the Simulation

Five panels and windows will appear when you start the simulation: the P&G Manager and Run Control panels, an AlphaSynapse panel beneath these controlling panels, a Voltage-vs-Time graph, and a Voltage-vs-Space graph.

### 3. The synaptic input
An AlphaSynapse panel controls the parameters of the single input to the dendritic tree. It can be reopened, if closed, by a button on the P&G Manager. The parameters are the same as in the similar panel in the Patch Postsynaptic Potential Tutorial with the addition of a slider that allows you move the location of the synapse to any position between the tip of the dendrite and the soma. The location of the synapse is specified beneath the slider as a fraction of the distance from the dendrite tip (1.0) to the soma (0.0).

### 4. The graphs

*a. The Voltage-vs-Time plot (upper right)*
This graph has recording electrodes at three locations:
- in the **middle (0.5) of the dendrite (blue trace)**
- in the **soma (red trace)**
- at the **(0.1) position on the axon (black trace)** (10% of the distance from soma to its end)

*b. The Voltage-vs-Space plot (lower right)*
The dimensions of the dendritic tree, soma, and axon comprising this x-axis are shown on the diagram above. The positions of the three recording electrodes are indicated on the x-axis by the color-coded arrowheads.

---

# Experiments and observations

## *Where is the site of spike initiation?*

### 1. Deliver a suprathreshold EPSP to the middle of the dendrite.
Note the default parameters in the AlphaSynapse panel, then click R&R. Observe where the impulse initiates when the synaptic input is at its default location in the middle of the dendrite.

**2. Study the movie and recorded traces.**
Using Reset, then "Continue for (ms)" steps of 0.2 ms, look carefully at the sequence of voltage changes in each domain of the neuron. Also look at each of the three recordings in the Voltage-vs-TIme plot.
- Question: Can you discern precisely where the impulse initiates and why?
- Question: Why is the action potential in the soma of reduced amplitude compared to that in the axon? Is this voltage change an action potential?
- Question: Why does the dendrite appear to give an action potential (or at least a voltage change similar to that in the soma) when it has been specified not to contain voltage-sensitive ion channels?

**3. For fun, watch impulse initiation and propagation in pseudocolor.**
As another way of visualizing the voltage distribution and propagation along the axon, you can click the "Pseudocolor Plot" button in the P&G Manager. A special window comes up and overlies the Voltage-vs-Time window. On the left of this Pseudocolor plotting window is a scale of voltage as a series of colors from deep red to yellow; the x-axis is distance, as in the Voltage-vs-Space window. Click R&R to see the impulse propagate as a wave of color.

## Can the site of impulse initiation be altered by parameter changes?

Could the default settings have been selected to cause initiation in the axon? Was this a setup?

**1. Change the synaptic strength (conductance).**
- Double, then redouble the synaptic conductance (gmax).
- Question: Has the site of initiation changed with synaptic strength?

**2. Change the location of the synapse on the dendrite.**
- Return the synaptic conductance setting to its default value.
- In the AlphaSynapse Manager, move the synapse first to the distal and then to the proximal end of the dendrite and run the simulation in each location.
- Question: Has the site of initiation changed with synaptic location?

**3. Change the morphology of the soma.**
- Restore the default synaptic location In the synaptic panel.
- Press the "Cell Parameters" button to bring up menus for all three regions of the cell.
- In the Soma Parameters panel, make the following changes and run the simulations:
  - Change the soma's diameter over the range from 1/2 to 4 times its default value of 200 μm.
  - Double and halve the soma's length from its default value of 200 μm.
- Question: Can you explain your results when you change the soma morphology?

### 4. Change the diameter of the axon.

- In the Soma Parameters panel, restore the soma's diameter and length to their default values of 200 μm.
- In the Axon Parameters panel:
  - Change the axon's diameter over the range from 1/2 to 4 times its default value of 15 μm.
  - As you increase the axon's diameter, you may want to increase the synaptic strength (gmax) to keep the time of spike generation approximately the same.
- Question: Has the site of initiation or the relative timing of the spike in the soma and axon <u>changed</u> with axon diameter?

### 5. Change the ion channel densities.

In freeze-fracture electron micrographs taken at high power, there appears to be an increase in the number of particles seen in the membrane of the initial segment of axons relative to other regions. There is now experimental evidence that these particles may be channels. You can explore the effect of changes in axonal channel densities here.

- Restore (if altered) the synaptic strength to its default value.
- In the Axon Parameter panel:
  - Restore the axon's diameter to its default value of 15 μm.
  - Run the simulation and employ the "Keep Lines" option in the Voltage-vs-Time window.
  - Halve, then double the densities (conductances) of both the Na and K channels.
  - Leave the Na channel density doubled but return the K density to its default value to see the effect of doubling the Na channel alone.
- Question: Have changes in ion channel densities <u>affected</u> the site of impulse initiation or the relative timing of the spike in the soma and axon?

---

## Summary

A neuron's morphology *per se* causes the spike to initiate in the region where the soma tapers into an axon. The probability of this region being the site of initiation can be enhanced by additional ion channels at this location (especially Na channels). There is mounting evidence that the initial segments of axons of some neurons have just such an extra dose of Na channels.

# Synaptic Integration Tutorial

The neuron you will simulate here is a simplified representation of the spinal motoneuron of the cat. NEURON is capable of handling the full set of morphological details of a real neuron: its structure, channel types, channel locations, and densities, and the synaptic locations throughout the full dendritic tree. In this tutorial, however, we will restrict these parameters to a <u>manageable subset</u>.

Even with this reduced parameter set, you may begin to feel overwhelmed by the complexity introduced by the possibilities. There is much to be learned simply by changing the strengths, locations, reversal potentials, and onset times of the three synaptic inputs on the dendrites. We hope these tutorials will raise your sensitivity to oversimplifications.

## Geometry of the simplified motoneuron

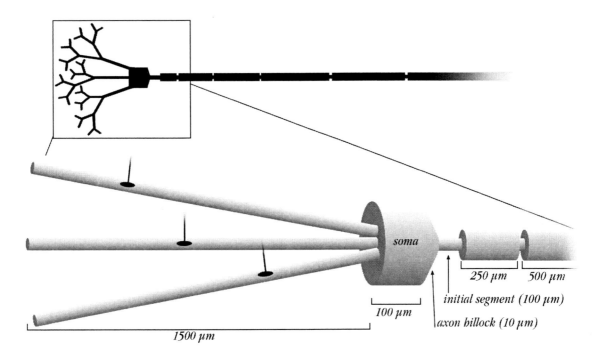

### The sections of the neuron

1. **The dendritic tree**
   The dendritic tree is now represented by three dendrites 1.5 mm in length and 24 µm in diameter, each with synaptic inputs whose location, strength, and timing may be controlled.

2. **The soma and axon hillock**
   The soma is represented by a cylinder 100 µm by 100 µm; the axon hillock is the 10 µm region where the soma tapers from 100 µm at the soma to the 10 µm diameter of the specialized region called the "initial segment" and also of the axon.

**3. The initial segment and the axon**
Between the axon hillock and the axon is the initial segment of bare axon,
100 μm in length and 10 μm in diameter. The rest of the axon, 5 mm in length, is
myelinated (rather than bare as in the Site of Impulse Initiation Tutorial). You will
explore whether the presence of myelin is important in determining the site of
impulse initiation.

## *Location and densities of channels in the neuron*

For simplicity, only Hodgkin-Huxley channels have been placed in the motoneuron. The
densities of these channels in each component structure have been chosen to match
available experimental observations on the spinal motoneuron of the cat. Recent evidence
indicates that the dendrites of several types of CNS neurons, including motoneurons, may
have voltage-sensitive channels but for didactic purposes the dendrites in this tutorial are
passive.

# Goals of this tutorial:

- To compare the site of impulse initiation in this neuron, whose axon is myelinated,
  with that in the unmyelinated axon in the Site of Impulse Initiation Tutorial.
- To explore how impulse initiation depends on each of the following:
  - the timing of the synaptic inputs on each of three dendrites
  - the location of these synapses
  - the temporal pattern in a train of impulses arriving at a single synapse
  - the temporal pattern, strength, location and reversal potential of the synaptic
    inputs on each of the three dendrites

# Set up for the tutorial.

**1. Assumptions**
This tutorial assumes that you are incredibly facile with NeuroLab manipulations,
even making new right-mouse-button discoveries on your own. It also assumes
that you have done the experiments in the Patch Postsynaptic Potential Tutorial
and the Site of Impulse Initiation Tutorial.

**2. Starting the tutorial**

### Start the Simulation

Typical panels and windows will appear when you start the simulation. Their spe-
cializations for this tutorial are described below.

**3. The synaptic input**
Each of the experiments in this tutorial has a special, dedicated AlphaSynapse
panel. Each of these panels is called up in the same position (lower-left portion of
the monitor) by its button in the P&G Manager at the appropriate moment in the

tutorial. The first panel is controlled by the "Onsets, Locations" button (see also title in blue bar at the top of the panel) but comes up automatically when you click Start the Simulation.

The Onsets, Locations panel that first comes up allows you to change certain variables of the synapses on the three dendrites:
- **The conductance**, which in this menu is the same for all inputs
- **The onset** of each synaptic potential, which can be adjusted for each synapse separately
- **The location** of each synapse on the dendrite, adjustable by a slider for each synapse as shown in the diagram

The text line just below the three sliders shows, from left to right, the location of each synapse as a fraction of the distance from the tip (1) of the dendrite to the soma (0 .

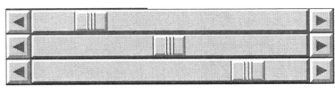

<END Syn 0-2 Locations    0.823    0.500    0.177    SOMA>

### 4. The graphs

   *a. The Voltage-vs-Time graph (upper right)*

   Four electrodes record the voltage change in the neuron in these color-coded positions:
   - in the soma (thicker trace)
   - at node [0] (0.35 mm from soma)
   - at node [1] (1.1 mm from soma)
   - at node [5] (5.1 mm from soma)

   *b. The Voltage-vs-Space graph (lower right)*

   The dimensions of the dendritic tree, soma, and 5-mm-long axon comprising this x-axis are shown on the diagram above. The positions of the four recording electrodes are indicated on the axis by the color-coded arrowheads.

## Experiments and observations

### *Generate EPSPs in the dendrites and observe where the spike initiates.*

1. **Deliver three EPSPs 0.5 ms apart at three dendritic locations.**
   In this first simulation, note the staggered locations of the inputs (EPSPs) to the three dendrites by checking the positions on the sliders. They occur in sequence—at 0 ms for the input onto the black dendrite, 0.5 ms for the input onto the red dendrite, and 1 ms for the input onto the blue dendrite. Now press R&R to see <u>where the spike is initiated</u> with these inputs.

**2. Examine the voltage change in the initial segment and axon.**

For a paused-action view of the sequence of events, advance the movie with the"Reset" and "Continue for (ms)" buttons in increments of 0.5 ms or another value of your choice. Notice especially the voltage change in the initial segment. Why is it so steep?

**3. Examine the voltage change in the soma and dendrites.**

The impulse, initiated in the axon, backpropagates into the soma, its peak amplitude progressively decreasing as it does so from node[5] to node[0]. The soma struggles even to depolarize up to 0 mV. Backpropagation of the depolarization continues into the dendritic tree where most of each dendrite is depolarized to about -40 mV.

• Question: Recent experiments reveal that spikes can backpropagate into the dendritic tree in a number of CNS neurons. What effect will this dendritic depolarization have on subsequent synaptic inputs on any dendrite? Click here for a discussion of the question.

**4. Reverse the timing or locations of the EPSPs so that the EPSP near the soma comes first.**

Deliver the first EPSP near the soma and the last near a dendritic tip. You can do this either by simply reversing the values of the onset times for Syn(apse) 0 and Syn(apse) 2, leaving their locations the same, or by reversing synapse locations as shown below. The EPSP occurring first will now be closest to the cell soma. Will this sequence of EPSPs cause an action potential to initiate? Click here for a discussion of this experiment. Experiment with moving the location of the most distal synapse to see how its position affects your results.

Original positions:

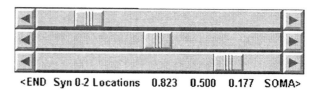

&lt;END  Syn 0-2 Locations   0.823   0.500   0.177   SOMA&gt;

Reversed positions:

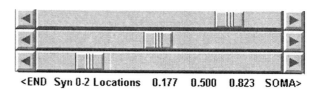

&lt;END  Syn 0-2 Locations   0.177   0.500   0.823   SOMA&gt;

## *Deliver a train of synaptic inputs at a single synaptic location.*

The temporal pattern of impulses in a train arriving at a synapse is thought to be important in determining whether or not a neuron fires. In this tutorial you can arrange to have the three EPSPs arrive at one synapse on one dendrite, simulating the first three impulses of a train.

**1. Bring up a new AlphaSynapse menu.**

Click the "Train, Single Location" button in the Panel Manager. There is no need to close the previous menu; the new one will replace it.

### Trains in NEURON

The simulator views a train as three synapses at the same location (indicated by the slider position), in this case on the red dendrite. The onset of each EPSP is set independently.

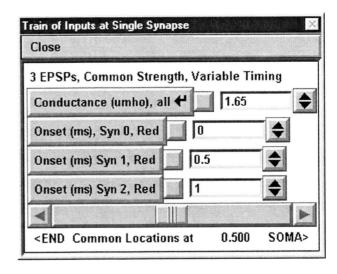

**2. Deliver the train.**

Run the simulation with the default parameters and watch carefully the movie of the three EPSPs. The conductance increases have been made identical for each EPSP to keep this experiment from becoming too complex. If you have completed the Patch Postsynaptic Potential Tutorial, you should be able to explain why the three EPSPs do not add linearly even though the conductance changes are the same.

**3. Are three EPSPs more effective when they occur together or when they are in a train?**

*a. Determine the threshold conductance for the three EPSPs in the train at the 0.5 location.*

The fastest way to search for the threshold conductance is to close the Voltage-vs-Space graph, because movies slow down the simulations, and simply watch the Voltage-vs-Time graph to detect when an action potential is generated. It is also faster to use the arrows in the spinner box to make your changes (rather than the field editor), because a simulation will automatically start when you make the change. (Reminder: Clicking the right mouse button on the UP arrow in the spinner box will allow you to select the decade in which the arrows of the box make changes.)

*b. Now set all three onset times to 0 ms and determine the threshold conductance again.*

Explain your results. As an advanced exercise to test your explanation you can experiment with the timing of the EPSPs in the train. What is the lowest synaptic conductance that you can achieve by playing with the interspike intervals?

4. **How does threshold change with position of the synapse on the dendrite for a train?** Find the threshold synaptic conductance as a function of the position of the synapse along the dendrite for the EPSPs occuring in the train and also occuring simultaneously. <u>What do these relations tell you</u> about voltage spread in a passive dendritic arbor?

5. **How does threshold change with position of the synapse on the dendrite for EPSPs occuring simultaneously?**

## How does inhibitory synaptic input (an IPSP) at one synapse affect voltage spread and impulse initiation?

1. **Press the "All Synaptic Variables" button.**
   The AlphaSynapse menu that is called up allows you to vary four parameters for each synapse:

   - conductance
   - onset
   - location
   - reversal potential

2. **Make one synaptic input an IPSP.**
   Set the reversal potential of one of the synapses to a value more negative than the resting potential. Run the simulation. You have the tools to do a large number of experiments with this neuron. For example, you can test parameters that might change the effectiveness of the IPSP in blocking impulses generated by one or two EPSPs. (You can set a conductance to zero to remove an input.) For the IPSP, explore the effect of changing:

   - its onset time with respect to EPSPs
   - its location
   - its conductance change
   - the value of its reversal potential

3. **Not satisfied? Change the temperature.**
   If you find these experiments insufficiently challenging, you can investigate how temperature affects spike generation for the whole parameter set!

# Presynaptic Terminal Tutorial

In the Synaptic Integration Tutorial, the impulse, generated in the myelinated axon of a motoneuron, sets off at high speed toward its destination. After an interval of a few milliseconds, we pick up its journey here as it flies into the nerve terminal on the muscle fiber. What happens to an action potential in a myelinated axon as it invades the unmyelinated presynaptic terminal?

## Geometry of the presynaptic terminal in this tutorial

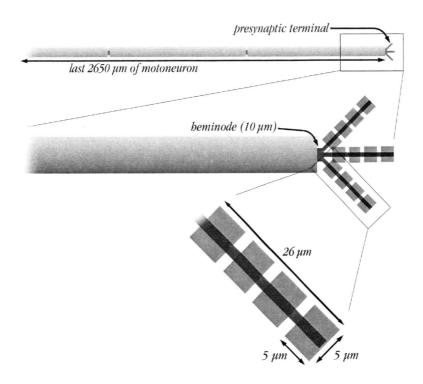

### *The plaque-type neuromuscular junction*

In vertebrates, there are two basic types of neuromuscular junction on skeletal muscle. The bouton-type neuromuscular junction of the lizard has been chosen for these simulations.

### *Morphology of the terminal region*

For these simulations, a myelinated axon 10,020 μm long (myelinated regions totaling 10,000 μm plus 10 nodes totaling 20 μm) loses its myelin in a region called the "heminode" (10 μm long). The axon then trifucates into a terminal arborization comprising three equivalent branches, each 28 μm in total length. Each branch contains four boutons (5 μm in both diameter and length), spaced evenly along the branch 2 μm apart. The branches are numbered by NEURON as branch[0], [1], and [2] from top to bottom, but we will always be concerned with the middle branch[1]. This stylized morphology is based on published light and electron microscopic observations.

In this tutorial you will examine first the voltage along the axon, heminode and middle branch of the arbor (an overall length of 10,056 μm), then the current patterns along the middle branch.

### Channel densities

Channel density assignments are based on a published model.

- The Na channel density (Na conductance) at the nodes in the axon is the same value as in the Myelinated Axon Tutorial, approximately ten times the density in squid axon.
- The Na conductance at the heminode is half of that at a node. This channel density declines further in the boutons but remains greater than that of squid axon membrane.
- The decreasing Na conductance is mirrored by an increasing K conductance until the last bouton.

## Goals of this tutorial:

- To observe the voltages at the termination of the myelin (the heminode) and at the boutons in the presynaptic arbor as they are invaded by an action potential
- To experiment with temperature, internode length, and Na channel density to see how each parameter affects invasion of the terminal arbor
- To observe currents throughout the arbor and compare your simulations with experimental observations
- To experiment with the parameters of each bouton, observing how changes in bouton geometry or conductances affect the currents in the terminal

## Set up for the tutorial.

### 1. Assumptions
By now you should be totally at ease with the manipulations of NeuroLab. This tutorial builds on the Level I Axon and Nonuniform Axon tutorials, especially the Partial Demyelination Tutorial.

### 2. Starting the tutorial

**Start the Simulation**

The following panels and windows will appear when you start the simulation:
- A P&G Manager and a Run Control panel
- A Stimulus Control panel
- A Voltage-vs-Space plotting window

### 3. The stimulating electrode
The Stimulus Control panel shows that the axon is stimulated at the far left. You will not be asked to make changes in the stimulus location or settings but the panel must be open in order to stimulate the axon.

## 4. The graphs
A Voltage-vs-Space graph will display the voltage as a function of length of the entire axon plus the central branch of the terminal. The x-axis is only 10,000 μm long but the entire axon is 10,056 mm long so the voltage trace will actually extend beyond this axis by 56 μm.

# Experiments and observations

### *View movies of an impulse propagating along the axon and into the nerve terminal.*

### 1. View the action potential propagating into the central terminal branch.
When you run the simulation, the cell will be stimulated at its left end (mm = 0). Watch the right end of the voltage trace: here the tiny central terminal arbor (28 μm) follows the voltage in the axon almost but not quite, so that the voltage suddenly droops in this region.

### 2. Study the impulse in the axon by repeatedly pausing the movie.
For a slow motion view, press Reset, then, repeatedly, the "Continue for (ms)" button as usual. Doing this will allow you to notice the cusps in the voltage profile that occur at nodes. You can see that they appear at 1 mm intervals along the axon.

More accurate location of the nodes can be established by sliding the crosshairs along the voltage profile. Remember that there are 2 μm nodes between each internode.

### 3. Focus on the terminal arbor.
Bring up the Expanded Voltage-vs-Space graph. (It will overlie the Voltage-vs-Space graph.) The x-axis of this graph shows the final 256 μm of the axon including the 28 μm central terminal branch. When you run the simulation you will see the voltage decline like a staircase from bouton to bouton along the arbor. Again, using Stop or Continue for (ms) is useful. Use the crosshairs to determine in which portions of the branch the voltage is falling most steeply—in the boutons or in the processes between the boutons.

It takes a little while after pressing R&R for the impulse to reach this expanded end of the axon. If you try to speed the process by stimulating the axon nearer the terminal, you will notice that you need more current to set up an action potential in the axon. Can you guess why this is so? Hint: The left end of the axon (x = 0 mm) is <u>closed</u>.

### *Record voltage-vs-time traces at selected positions along the axon and terminal.*

### 1. Bring up the Voltage-vs-Time window.
This window will display recordings from seven (!) color-coded electrodes when you run the simulation. Although this array may seem daunting, when you run the

simulation you may appreciate what is happening more easily than you might think. The electrode positions are as follows:
- Four of the electrodes are at nodes 9, 7, 5, and 3 (with node numbers decreasing in the direction of the nerve terminal).
- The fifth electrode is located at the heminode.
- The sixth and seventh electrodes are in the central terminal branch (branch [1]), the blue electrode at the bouton nearest the axon (bouton[1] [0]) and the brown electrode at the third bouton in the line (bouton[1] [2]).

2. **Study the Voltage-vs-Time traces in the heminode and terminal branch.**
When the action potential hits the heminode and invades the nerve terminal, its amplitude is suddenly greater. The voltages throughout the terminal, while declining with distance from the heminode, are all greater than in the axon. <u>Why are these voltages larger?</u>.

---

## *Experiment with factors affecting invasion of the terminal.*

When the axon loses its myelin at the terminal region, the situation seems similar to that in the Partial Demyelination Tutorial where a <u>change in temperature</u> or in the <u>length of the internode</u> adjacent to the bare axon was critical in supporting impulse invasion of that region. What factors are important in invasion of the terminal?

1. **How does temperature affect invasion of the terminal?**
Raise and lower the temperature by, say, 5°C. Explain any changes you observe in the peak and shape of the action potentials; you should have considerable experience with temperature effects by now from earlier tutorials. Restore the temperature to its default value of 25°C before the next experiment.

2. **How important for invasion is the length of the last internode?**
In the <u>Partial Demyelination Tutorial</u> you observed that reducing the length of the internode next to the bare axon improved the chances that the impulse would propagate into this unmyelinated portion. How important is the length of the last internode at the nerve terminal?

   *a. Bring up the Internode Lengths panel.*

   Reduce the length of the last internode (Myelin[0]) dramatically, say, to 50 μm (the length actually observed at the lizard terminal). Since you are reducing the length of the axon by doing this, the voltage change in the terminal branch will happen earlier. But, beyond this, what is the effect of <u>changing the length of this internode</u> on the action potentials of the heminode and the boutons?

   *b. Return all parameters to their default settings.*

3. **How important for invasion is the Na channel density in the heminode?**
The heminode is the "half-node" joining the axon to the terminal arbor. Open the Heminode Parameters panel. Its Na conductance is 5 times the squid axon value. Must it be so large to enable invasion? What happens to invasion of the terminal if you divide this value by 5, returning it to the "normal" (squid axon) value?

*a. You can check on the values of Na conductance in the boutons.*

Open the Bouton Parameters tray by pressing the "Bouton Parameters" button in the P&G panel. The Na conductances in the boutons are higher than normal in all of the boutons. If you set them all to the squid axon value of 0.12, what is the effect on invasion?

*b. Return all parameters to their default settings.*

## Observe the current patterns throughout the terminal arbor.

1. **Open the Terminal Currents graph.**
   In this window you can plot the currents at various locations in the nerve terminal, starting with the heminode and proceeding through the boutons going out the middle branch.
   - The label "**heminode**" identifies the current in the "half-node" joining the axon to the terminal arbor.
   - Bouton[1][0] identifies the current in the first bouton (bouton [0]) in the middle branch (the [1] branch). See the diagram for help! Again, we apologize that NEURON starts sequences with zero and not with 1. (Please remember that NEURON was designed as a research tool and uses standard mathematical nomenclature.)
   - Bouton[1][1] identifies the second bouton.
   - Bouton[1][2] identifies the third bouton.
   - Bouton[1][3] identifies the fourth bouton.

2. **Examine the currents throughout the arbor.**
   Run the simulation and interpret your traces. The only inward current is at the heminode; the remainder of the currents are outward.

   You can call up the Heminode Parameters panel and the Bouton Parameters tray of panels to check on the values of the ion conductances at each location to aid in interpretation.

3. **Compare the currents with published observations.**
   Compare the current traces with published <u>experimental observations</u>. Notice that the published records include a stimulus artifact not present in the simulations of the tutorial.

4. **Change the bouton conductances.**
   In the Bouton Parameters panel, test the sensitivity of these current patterns to changes in the ion conductances. Increase and decrease the Na conductance for each bouton twofold; then do the same for the K conductance.

   As you do this, you will gain an appreciation of how a change in one bouton can affect currents throughout the terminal. We hope you will begin to understand the difficulties in the <u>process of matching</u> simulated traces to experimentally observed traces.